The Power of Words in International Relations

Politics, Science, and the Environment
Peter M. Haas and Sheila Jasanoff, editors

Peter Dauvergne, *Shadows in the Forest: Japan and the Politics of Timber in Southeast Asia*

Peter Cebon, Urs Dahinden, Huw Davies, Dieter M. Imboden, and Carlo C. Jaeger, eds., *Views from the Alps: Regional Perspectives on Climate Change*

Clark C. Gibson, Margaret A. McKean, and Elinor Ostrom, eds., *People and Forests: Communities, Institutions, and Governance*

The Social Learning Group, *Learning to Manage Global Environmental Risks.* Volume 1: *A Comparative History of Social Responses to Climate Change, Ozone Depletion, and Acid Rain.* Volume 2: *A Functional Analysis of Social Responses to Climate Change, Ozone Depletion, and Acid Rain*

Clark Miller and Paul N. Edwards, eds., *Changing the Atmosphere: Expert Knowledge and Environmental Governance*

Craig W. Thomas, *Bureaucratic Landscapes: Interagency Cooperation and the Preservation of Biodiversity*

Nives Dolsak and Elinor Ostrom, eds., *The Commons in the New Millennium: Challenges and Adaptation*

Kenneth E. Wilkening, *Acid Rain Science and Politics in Japan: A History of Knowledge and Action Toward Sustainability*

Virginia M. Walsh, *Global Institutions and Social Knowledge: Generating Research at the Scripps Institution and the Inter-American Tropical Tuna Commission, 1900s–1990s*

Sheila Jasanoff and Marybeth Long Martello, eds., *Earthly Politics: Local and Global in Environmental Governance*

Christopher Ansell and David Vogel, eds., *What's the Beef? The Contested Governance of European Food Safety*

Charlotte Epstein, *The Power of Words in International Relations: Birth of an Anti-Whaling Discourse*

The Power of Words in International Relations

Birth of an Anti-Whaling Discourse

Charlotte Epstein

The MIT Press
Cambridge, Massachusetts
London, England

For information about special quantity discounts, please email special_sales@mitpress.mit.edu

This book was set in Sabon on 3B2 by Asco Typesetters, Hong Kong.
Printed on recycled paper and bound in the United States of America.

Library of Congress Cataloging-in-Publication Data

Epstein, Charlotte, 1975–
The power of words in international relations : birth of an anti-whaling discourse / Charlotte Epstein.
p. cm. — (Politics, science, and the environment)
Includes bibliographical references.
ISBN 978-0-262-05092-0 (hardcover : alk. paper) — ISBN 978-0-262-55069-7 (pbk. : alk. paper) 1. Whaling—History. 2. Whales—Conservation. 3. International relations. 4. Discourse analysis. I. Title.
SH383.E67 2008
333.95′95—dc22 2008007299

10 9 8 7 6 5 4 3 2 1

Contents

Preface

Let me make this absolutely clear right from the start: in no way am I advocating the killing of whales. What has puzzled me about these animals is the way in which, sometime in the middle of the twentieth century, we shifted dramatically from a world where killing whales was widespread and unquestioned to one where it is morally wrong and those who continue to do so are frowned upon. Yet, by that point, blue whales had been known to be endangered for over three decades, and this did not stop us from hunting them down all the harder. Was this really, then, about protecting whales? Or was this more about us humans and how we interact with one another? Or perhaps it was both, about how we relate to one another and to our natural environment.

Another important clarification is, I feel, necessary from the onset. The whaling issue is also a story about environmental nongovernmental organizations (NGOs). Indeed, it is their first global success story. It is always quite uncomfortable to critique the "good guys." This book should not be read as a condemnation of NGOs, whose broader political function is not under question here. It was they, after all, who brought home the realization that if we did not stop, there would be no more whales left in our seas. However, as with any other political actors, they cannot be spared the critical scrutiny they are sometimes loath to accept, especially since they often escape the more traditional systems of political accountability. Their powers can be quite extensive indeed, as this story will show, and as with any other powers, they cannot be left unchecked.

This leads me to my choice of discourses. First, I should specify that I am talking about words strung together in a discourse, not words on their own. There are a few methodological reasons why discourses on whales and whaling lent themselves especially well to a discursive approach, which are revealed in the following chapter. The point I want to make here is that too often discourse analyses tend to target discourses that belong to "other," distant people—no doubt because, not being immersed in them, the critical distance is easier to achieve. To me, it was important to take one of our own, one that *we* tend to reproduce automatically without pausing to consider what we are saying. I too, like most people of my generation, grew up with the conviction that whales needed to be saved from the cruel people who kept killing them. I remember how, as a child growing up inland, who had never actually seen a whale, I would stick "Save the whales" logos to our bathroom walls and muse about these magical creatures while wallowing in the tub. The picture was, as I found out, a little more nuanced than the loud logos would let on.

Another key reason for my choice is that whales are useful for dispelling the still widespread misperception that studies that take discourses as their main focus cannot handle material reality—that they lose themselves in endless deconstructions where reality soon dissolves into thin air. More broadly, the enquiry into the social construction of reality does not deny that there is a reality "out there." Whales are indeed very big, very real, and very far out there. Moreover, our relationship to them is not just about words but rather about very material, indeed sometimes very bloody practices. Casting the lens upon real, natural creatures is also especially useful for emphasizing the task of "de-naturalising the taken-for-granted" (Weldes et al. 1999, 19) that drives the critical examination of the social processes by which we construct the world we live in. Whales are no doubt natural creatures. The way we relate to them, on the other hand, is socially constructed—so much so that, historically, we have been able to find ourselves in two radically opposite relationships with them: one that was all about killing them, and the other, all about saving them. The former is useful for bringing into relief the latter, the one we now naturally take to be the "right" way of envisaging whales. The key insight that brought about the reflexive turn in the social sciences is the realization, not only that the social world is constructed but, consequently, that it could have been constructed otherwise. What whaling draws into relief is that *it really was* construed in a very different way.

Finally, I would like to say a word about a specific relationship between theory and practice that I attempt to develop in this book. Let me start with an anecdote: one of my reviewers queried whether the book was about discourse or about whaling. The answer is both, in equal measures. Whales/whaling is simultaneously an object of analysis in its own right and a lens for examining key theoretical concerns related to the use of discourse in analyzing international relations. Consequently, the book breaks the habitual separation where a "theory chapter" caps the "case study chapters." Rather, the theory is woven into the analysis of the case itself. Every chapter considers an aspect of the case that speaks to a specific theoretical concern. Each one begins by identifying the theoretical issues drawn out by the case and by positioning them in relation to the broader literatures. This choice beckons your patience as a reader. While you are welcome to travel along the chapters and select specific aspects of either the case or the theory, like as many dishes on a buffet, I would like to invite you instead to sit down to the full course meal, which I hope to be a far more satisfying experience.

At this stage, I will simply state the two axes around which the book is laid out. The first is horizontal and temporal—a before/after axis. The second is a vertical, levels-of-analysis axis. The book is structured around a rupture, between a past whaling world and a current anti-whaling world. This rift articulates the passage from part I to part II of the book as a whole, but it also runs through many of the chapters themselves. The focus on discourse—what the social actors say—enables the analysis to scale up and down the levels of analysis, from the individual to the state, and vice versa.

Research for this book began as a PhD project at the University of Cambridge, and I would like to thank my supervisors, the late Nick Sinclair-Brown and James Mayall, as well as my friends in Cambridge for some lovely times there. I had the opportunity to observe the international politics of whaling "in vivo," as it were, by attending two annual meetings of the International Whaling Commission (IWC) in London in 2001 and in Berlin in 2003. I also spent some time in with the whalers in Lofoten off the coast of Northern Norway in the summer of 2003. I would like to thank my hosts in Lofoten, Laila Jusne and Rune Frovick. I am also thankful to the helpful staff of the IWC Secretariat, the many scientists, NGO officials, and members of state delegations I was able to interview. I acknowledge the financial support of the British Council (for a Chevening scholarship), the British Academy's Arts and Humanities Research Board, and the Cambridge European Trust as well as that of my college, Sidney Sussex. A scholarship from the Georges Lurcy Foundation enabled me to finish the PhD as a visiting scholar at the University of California, Berkeley, and I am grateful to my friends there for distracting me with the joys of the Bay Area. Finally, I thank Sydney University's Research Institute for Humanities and Social Sciences for granting me a Writing Fellowship that bought me precious time to bring the book to a close in Sydney. On a more personal level, I thank my mother, Anne-Marie Epstein, for believing in the project well before I ever began to, my father, Marc Epstein, who, in never remembering its title, made me realize the need to change it, my sister Sophie Epstein, for feeding me endless jokes about whales (and there are many), Evatt Hawkes, a "hard" scientist readily convinced of the power of words, for his unwavering support, and Mark Blyth, for patiently putting up with my arguing against every one of his suggestions for improving the manuscript.

Acronyms and Abbreviations

ACMMR	Advisory Committee on Marine Mammal Research
AEWC	Alaskan Eskimo Whaling Commission
BWU	Blue Whale Unit
CIA	Central Intelligence Agency
CITES	Convention on International Trade in Endangered Species
CLA	Catch Limit Algorithm
EEZ	Exclusive Economic Zone
ESA	Endangered Species Act
FAO	Food and Agriculture Organization
FIDS	Falkland Islands Dependencies Survey
FoE	Friends of the Earth
GATT	General Agreement on Tariffs and Trade
GDP	Gross domestic product
HNA	High North Alliance
ICR	Institute for Cetacean Research
ICRW	International Convention for the Regulation of Whaling
IDCR	International Decade for Cetacean Research
IFAW	International Fund for Animal Welfare
IUCN	International Union for the Conservation of Nature
IWC	International Whaling Commission
JARPA	Japan's Special Permit Research on Minke Whales in the Antarctic
MMPA	Marine Mammal Protection Act (U.S.)

MSY	Maximum Sustainable Yield
NAMMCO	North Atlantic Marine Mammal Commission
NGO	Nongovernmental organization
NIO	National Institute of Oceanography
NMP	New Management Procedure
OECD	Organization of Economic Cooperation and Development
RMP	Revised Management Procedure
RMS	Revised Management Scheme
SC	Scientific Committee
SOWER	Southern Ocean Whale and Ecosystem Research
UN	United Nations
UNCED	United Nations Conference on Environment and Development
UNCLOS	United Nations Convention on the Law of the Sea
UNEP	United Nations Environment Programme
UNESCO	United Nations Educational, Scientific and Cultural Organization
WCW	World Council of Whalers
WWF	World Wildlife Fund

The Power of Words in International Relations

1

Making Meaning Matter in International Relations

Well into the second half of the twentieth century the world was largely a whaling world. Whales comprised a strategic resource, a key raw material, a fuel, and a food. Whaling was just as important to us then as the oil industry is today, with the "baleens" providing us the equivalent of plastic, and whale oil lighting the streets of New York or London. Consequently, whales were ferociously hunted down, to the point where there were dangerously few left in the seas, and whaling itself became uneconomical. At the same time, other substitutes (notably petroleum and plastic) increasingly replaced the main uses of whale parts. Both people and states thus turned away from whaling. In fact, whales point to one of the most dramatic cases of complete turnabout with regard to a natural resource, and a fundamental restructuring of the resource base of our economies. Whale oil constitutes the only form of energy that our societies both centrally depended upon and turned away from completely. At a time when the reliance on oil raises increasing questions, this in itself is food for thought.

If that were the end of the story, the whales would have simply gone their way, and so would we. Yet the story did not stop there. What eludes an account that focuses exclusively on the configuration of material interests is why states continue to *care* so much about whales.[1] In 1946 twelve whaling states set up the International Whaling Commission (IWC) to endeavor to contain a ruthless and self-destructive trade. As whaling slowly petered out, by such materialist account, states should have lost interest in the international organization created to regulate it; at least those who no longer whaled should have. Yet states stayed. In fact, more states joined—and more yet, many more, including many who had never had anything to do with whaling, even some landlocked states, such that the IWC membership today is over six times larger than it was in the heyday of whaling and covers almost half that of the United Nations (UN). Since 1982, commercial whaling has been precluded under an international moratorium upheld by a majority of states at the IWC, and it is widely frowned upon. How did the course of whaling matters alter so suddenly and so completely? States' turning to save-the-whale policies can hardly be explained by security or economic interests or by any other material factors traditionally relied upon in political science to account for state actions. The main argument of this book is that this change was brought about by a *powerful discourse*. And so I begin by considering these two terms in turn.

Powerful Discourses

Discourse confers meanings to social and physical realities. It is through discourse that individuals, societies, and states make sense of themselves, of their ways of living, and of the world around them. *A discourse* is a cohesive ensemble of ideas, concepts, and categorizations about a specific object that frame that object in a certain way and, therefore, delimit the possibilities for action in relation to it. It is a structured yet open and dynamic entity. This book is concerned with the discourses about whales, that is, ways of knowing, envisaging, and talking about whales that determine what we do to them. In effect, it is concerned with two discourses, one geared toward killing whales and the other toward saving them, and with how the latter superseded the former in the second half of the twentieth century. Schematically, studying, perceiving, and writing about whales as an oil resource or as a raw material makes no sense in a society that no longer whales and that sees whales as endangered species. Discourses are inherently social phenomena. They are what bind individuals together and enable them to engage, interact, and function socially. There would be no society without discourse. Running through the social fabric, they are like the lifeblood of social formations.

A *powerful* discourse is, quite simply, one that makes a difference. The rise of the anti-whaling discourse delegitimized a hitherto normal and widespread practice at the global level. The effects of a discourse, that is to say, its power, are at the heart of this book. Methodologically, a powerful discourse, the anti-whaling discourse, provides an applied entry point into the theoretical question of the power of discourse. At the same time, as Michael Barnett and Raymond Duvall (2005) pointed out, the question of power has forcefully returned to the forefront of the study of politics since 2001.[2] This book partakes in the broader effort to reopen the question of power. For, as they also underlined, most attempts to resurrect the concept in the discipline of international relations generally fall back to "realist" conceptions of the state's ability to control and coerce—in short a statist, top-down understanding of power (see Baldwin 2006 for a good example). This reduction of power to its physical and manipulative dimensions overlooks the generative, facilitative, strategic aspects of power operating from the bottom up, in short, a *productive* power that constitutes the very meanings and social relations it regulates (Barnett and Duvall 2005, Goverde et al. 2000, Litfin 1994, Clegg 1989, Fowler 1985). Locating such power in the discourses themselves was, in turn, made possible by Michel Foucault's broader reexamination of the nature of modern power. Foucault identified a power that no longer operates on the model of the premodern sovereign exerting its will from above and without. Rather, modern power is immersed in the social body; it has shifted from its head to its arteries, to string out the classical analogy of the body politic. Modern power circulates through the social order, and what it produces (and reproduces) is the capitalist social order itself, both discursively, through disciplinary norms, and through the subjectivities it creates.

With Foucault, power is no longer a quality, an attribute, or a capacity of the subject (individual or state). His key contribution, as Gilles Deleuze pointed out, was to undo the assimilation of power with property or appropriation. "Power is exerted rather than owned; it is not the acquired or preserved privilege of the dominant class, but the overall effect of its strategic positioning" (Deleuze 1986, 32–33, my translation). Foucault revealed the fundamental fluidity of power: "power passes through individuals. It is not applied to them" (Foucault 2003, 29).[3] This reconceptualization has two major consequences for appraising power. First, it displaces the analytical focus to power *relations* rather than power per se. Second, it depersonalizes and, consequently, also *de-essentializes* power. That is, it replaces the notion of power as a "thing" wielded by particular social agents with a relational understanding of power, lodged within the discourses. Discursive power is not a fungible entity, yet it has very real effects. This type of power displaces the perspective of enquiry: the focus is not so much on what power *is* (its essence) but on *what it does* (Foucault 2003; see also Guzzini 2008). Henceforth the question of power becomes foremost an applied question (Foucault 1980; see also Weldes et al. 1999, 10; Weldes 1999; Edkins

1999; Lynn Doty 1996; Weldes and Saco 1996; Campbell 1992, 1994). This perspective entails a commitment to a situated research (Haraway 1991) that starts from a particular set of social relations within particular "regimes of practice" (Foucault 1991, 75) and works from the ground up, progressively unearthing power's particular modes of exertion within it. In other words, and centrally here, social relations are both *simultaneously* the locus of power and the site for the production of meaning. My main argument in this book is that these two key elements of social life, power and meaning, are fundamentally intertwined, and much more so than has tended to be recognized in the study of international politics.

Discourses as Signifying Practices

The Coconstitution of Discursive and Material Practices

Put simply, discourses are sense-making practices. We string words together into sentences to make sense of the world around us, both to ourselves and to others. The focus on discourse as practice obtains in the type of explanation that has not causality but *meaning* as its main focus. In one of the founding moves of the social sciences, Max Weber famously divided all studies of social phenomena into those concerned with *explaining* them, which seek to uncover the causal laws governing positive facts and operate around a clear-cut distinction between the subject and object of analysis, and those concerned with *understanding* them, which have taken the turn toward meaning (Hollis and Smith 1990). The latter, reflexive turn has cast the focus back onto the social processes by which we know and construct the world we live in, thereby paving the way for analyses broadly concerned with "the social construction of," to which this books belongs. While useful to legitimizing both of these ultimately "irreconcilable stories" (Hollis and Smith 1990, 215), this distinction is coming under increasingly critical examination from within the reflexive turn (Parsons 2007, 111–112; Hansen 2006; Bially Mattern 2005). It precludes apprehending "meaning" as a cause of social action and as a factor of change and continuity, thereby undermining its explanatory purchase. The point here is not to salvage the language of causality in the study of meaning but rather to clear the grounds for establishing that the discursive approach I propose here does away with the distinction between explaining and understanding.

Discourses are thus the focus for a type of analysis concerned with meaning, that is, an explanatory mode centered on the construction of meaning. Consequently, a key concern is with identifying *where meaning is produced.* Discourses interest us not for their own sake but only insofar as they comprise sites for the making of meaning. However, material practices too constitute loci where meanings are produced. Whaling is a very concrete, material practice. It is also the repository of a whole host of meanings pinned upon the whale (for example, as source of raw material, as a food, or as a fiendish beast) that are reproduced every time a whaling expedition sets out. Discourses and material practices are thus tightly bound up and mutually constitutive (see also Pouillot 2007), and a discursive study is centrally concerned with any kind of practice implicated in the making of meaning. Insofar as ways of consuming the whales (or not) serve to reproduce the particular sets of significations associated with them, consumption constitutes an important signifying practice, and it is central to the analysis here. Words are examined insofar as they signify, that is, insofar as they constitute *signs.* Hence any type of sign, written or oral, visual or auditive, may qualify. Concretely, the analytical material for this book includes words, actions, music, and centrally in the case of whales, images.

The coconstitution of discursive and material practices moves the debate beyond a dichotomy carried over, beyond Max Weber, from the old divide between ideational and materialist lines of explanation, which also surfaced in international relations' founding disciplinary debate (pitting "realists" and "idealists"; Hollis and Smith 1990). This divide, in turn, is fundamentally rooted in a Western philosophy of *essences.* Here is not the place to engage in a discussion about the philosophical merits of the materialist/ideational divide.[4] The point is simply that when one starts from concrete, real-life practices, the separation between the ideational or the discursive and the material collapses insofar as what is said about whales is intimately tied to what is done with them, and neither is possible without meaning.

The connections between "doing" and "saying" have been unpacked from many different directions. Speech-act theory, in the wake of John Austin (1962) and John Searle (1969), showed how, in speaking certain words (such as "I do"), we were performing certain deeds (getting married). Ludwig Wittgenstein's (1969) "language games" cast the analysis of language itself upon concrete observations of the ways in which meanings are produced and modified in specific social contexts. In his wake, any serious examination of *language-in-use*—another possible definition of discourse (Dijk 1985a)—therefore requires observing how it actually *is* used in everyday practices—for example, those that involve whales. Laclau and Mouffe (1985) did away with the binary altogether by taking the logical step beyond Foucault, who ambiguously retained the dichotomy between the discursive and the nondiscursive (Howarth 2000, 104). Since discourses have very real, material effects, as indeed Foucault had shown, it simply makes no sense to consider them as "immaterial." Meanwhile, from the sociological perspective, Bourdieu's analysis of *praxis* emphasized the unity of thought and action, as indeed did the title of one of his major works, *Le Sens Pratique* (1980), which highlights both the meaning and direction inherent in social practices.[5]

To summarize the points made so far, discourses comprise sense-making practices that regulate what we do with, for example, whales, by pinning certain meanings onto them (a lubricating fuel or a magnificent and rare creature). Put simply, discourses "do" two things of concern here. First, they constitute a "space of objects" (Milliken 1999, 233). They render real things, such as whales, meaningful to us in particular ways. This space of meaningful objects is the space of a particular discourse, and what constitutes it as a bounded structure—as a discourse about whales, as opposed to a discourse about something else. Second, discourses constitute the identities of social actors, by carving out particular *subject-positions*, that is, sites from which social actors can speak, as the I/we of a discourse (for example, as an anti-whaler). In what follows these two key dimensions of discursive productivity are further developed.

The Production of Meaningful Objects

Toward a Relational Understanding of Meaning

Along with the discourse/practice binary, a more fundamental distinction is dissolved within the reflexive turn, between the world and the word, between an objective reality "out there" and the subjective world of speech and thought, such that the former would be more or less accurately reflected in the latter. The word is not the mirror of the world, because meaning is neither "innate" nor "fixed" once and for all. The central tenet of constructivism, namely, that the social world, unlike the natural world, is not "given" but rather socially construed, has been made possible by the fundamental shearing of the relationship between the word and the object "out there" that underlies all correspondence theories of the world. This, in turn, was ushered in by a closer understanding of language, or rather the processes by which we make sense of the world around us. Ferdinand de Saussure's analysis of *signification* had shown that the relationship between the word and the object, or the *signifier* and the *signified*, far from being "innate" or "automatic" is purely arbitrary, since different languages each feature their own sign for the same object. This makes several critical moves possible. First, meaning is not inherent but contingent and always in the making, since words do not "contain" meanings as real things. In fact, words are inherently *empty*. For meaning is not a thing in itself, a positive entity or essence. Not only do words have no meaning in isolation but their meaning is both yielded and exhausted by the play of differences between them. For example, the meaning of "hot" is given by the contrast with "cold," and vice versa, and on its own the sound of the phoneme "hot" does not trigger any meaningful associations in the mind of someone who does not speak English. Meaning thus emerges not from an inherent relationship of the word and the object, or between the signifier and signified, but from *a contingent relationship between the signifiers* (or signs).[6] Hence what "fills" a word or signifier with meaning—what renders it "meaning-full"—is for it to be set into relations with other signifiers within a discourse. Discourses are the *articulatory practices* that create these relations (Howarth 2000, 1995) and, therefore, meaning itself.

Yet a rift has opened up within the scholarship broadly concerned with the "social construction of." Many constructivists in the field of international relations sought to "seize the middle ground" (Adler 2006; see also Guzzini 2000; Wendt 1999, 1992, 1987; Checkel 1998) and retain the link to the physical world "out there" by epistemological recourse to "scientific realism" (Wendt 1987, 1992). As part of broader attempts to "rescue the exploration of identity from the postmodernists" (Checkel 1998, 325), this move was perhaps strategically necessary to be able to continue to claim the all-powerful mantles of "science" and "realism" by asserting that it, too, was studying the real world in proper scientific fashion and should therefore not be dismissed as fiction. As this book will endeavor to demonstrate, clinging on to this link is not a precondition to being able to go out there and study either identity or the world as it really is. In other words, it is not only unnecessary methodologically but it amounts to collapsing back the very distinction (between the social and the natural world) that had opened up the space for constructivist approaches in the first place.

As I hope the whales will drive home, there is no disagreement among approaches concerned with examining the "social construction of" that there really is a world out there. Once again, whales are very real and very much out there. The question is thus not whether material objects exist but how they become meaningful for us. With whales an additional conundrum is how did they come to hold such contradictory meanings. The productivity of discourse does not mean that material objects are physically brought into existence—a rather incongruous thought, at least so long as there are some flesh-and-blood whales left in the seas. Rather, these physical objects are brought into a system of meaningful relations. They are *linguistically* brought into existence, by being placed into relationships with other objects within a system of signifying differences.[7] The focus on discourse offers ways of apprehending the fundamental unfixity underlying the making of meaning, instead of evacuating it in favor of a nostalgic return to that long-lost link of the word to the world.

Toward a Discursive Understanding of Power

Constructivists' long, drawn-out efforts to rehabilitate "ideas" as explanatory factors in their own right have successfully moved the debate beyond the materialist stronghold in the study of international politics. They no doubt deserve all the credit for breaking the hold of the power–interest dyad in the explanation of social phenomena. However, because their strategy has amounted to wanting to "be part of the debate," at which they have been extremely successful—to the extent that they inaugurated the discipline's so-called Third Debate (Lapid 1989)—they have by the same token left the terms of the debate itself intact. They have, as a result, maintained, and even helped to reproduce, an essentialist epistemology, when it need not be, *for what they themselves set out to do*, as this book will endeavor to show. In other words, they have both successfully triangulated *and* preserved an essentializing debate. To the extent that, forcing the trait here only slightly, to acquire legitimacy new studies of international politics need only specify which of the three—power, interests, or ideas—they are positing as their independent variable (Checkel 1998). My argument here is that the terms of the debate are still set within a positivist framework that is inherently geared toward approaching *all* factors as material factors, or quasi-essences.

Two main problems follow from this. First, and not withstanding efforts to examine issues of authority and legitimacy, there has been a tendency within the scholarship interested in the "social construction of" to evacuate power, especially in the branch concerned with norms (Risse 2006; Checkel 2005, 1998; Joachim 2003; Klotz 2002; Josselin and Wallace 2001; Finnemore and Sikkink 1998; Keck and Sikkink 1998; Nadelmann 1990). Interests have, for their part, been recovered, to show how they are constructed by the ideas people have and the cultures to which they belong. Overlooked, as a result, are nonmaterial, discursive forms of power. I do not mean the benign power to bring about consensus, or to persuade, which is how power's productivity has tended to be operationalized in the analysis of norms. For "being persuaded" is not the same thing as having no choice but to talk (and act) in a certain way, because other ways of talking about the issue have been actively evacuated, a possibility which is made little room for in the emphasis on persuasion. Discursive power has been neglected, I argue, because insufficient attention has been paid to the construction of *meaning* rather than ideas or norms. If the world we live in is socially construed, it could have been construed otherwise; that much few constructivists would disagree with. However, it is not just that these social constructs are historically contingent. More fundamentally, their construction has *excluded* other possible sets of articulations or meanings. The prevailing of one particular social construct is an effect of power. It is this specific point of the process (of social construction) that is foregrounded in the discursive approach developed in this book, the moment where alternative constructs were evacuated. While contingency is given full consideration at the theoretical level in the constructivist literature, it is seldom mobilized into an applied analytics of power.[8]

Casting the focus upon shared meaning rather than ideas in people's heads draws out the fundamental unfixity and indeterminacy underlying its construction. Meaning emerges out of a process of determination that excludes other possible sets of articulations. In fact, a primary determination in the fixing of an object's meaning is the evacuation of *what it is not*: in the Spinozist formula (1674), logically "every determination is a negation." This, however, is not merely a semantic but a social process that centrally involves power relations. The fixing of meaning or the filling of the signifier is the outcome of a political struggle, which has foreclosed other possible meanings. The rise of a "hegemonic articulation" signals the victory of a particular configuration of meanings and social relations (Laclau and Mouffe 1985; Howarth 2000, 1995). Likewise, what is experienced as "common sense" is produced within specific historical contexts. It is born of a progressive sedimentation, in which particular configurations of meaning were sealed in, and others left out. The ultimate *product* of a powerful discourse is common sense, or *doxa*, to use an expression revitalized by Pierre Bourdieu, the habitual ways in which a society thinks. A "natural*ized*" discourse is one whose statements are experienced as "obvious," "true," and even "necessary." In other words, their historical contingency is evacuated, the fact that alternative meanings were ruled out, that, at one point, other significations could have obtained. At the outset of this process, the "evident" is powerful because it is unquestioned. Hence securing the domain of the "taken-for-granted" seals the moment of victory, the point at which a discourse becomes dominant, when its frames of thought and action become entrenched as the only possible ones. Thus routinized, these frames become modalities of social regulation (see also Mottier 2002; Torfing 1999, 2002; Shapiro 2002; Carver and Hyvärinen 1997; Faiclough 1992; Dijk 1985a, 1985b; Seidel 1985; Fowler 1985; Danet 1985).

This common sense is also the locus of what Bourdieu (1983, 1991, 1998, 2002c, 2003) analyzed as "symbolic domination," that is, a form of power that *does not need to coerce*, because it commands consent—in fact, operating at the level of meaning and social interactions, it works consent from within, for it sets the terms that make these interactions possible in the first place.[9] Uncovering the workings of this discursive, social power was one of his most important contributions to a relational understanding of power. Because it sets the terms of the debate, symbolic domination forecloses from the onset the possibility of any "serious deliberative argument" (Checkel 2005, 813) taking place. In fact, it evacuates the need for such argument to take place at all. Because the possibility of deliberating is entirely built on the presumption of an ideal communicative situation—two actors talking and listening to one another on a relatively level playing field or public space—in placing so much emphasis on deliberation the socialization literature assumes (without demonstrating) relatively undifferentiated positions of power and the discursive autonomy of the social actors engaging with one another. This presumption of equal positions of power is yet another way in which actual power relations are evacuated. What has been treated as persuasive authority may sometimes be nothing more than the power to impose norms without actually being seen to be doing so, by dictating the terms within which the deliberative argument will take place.[10]

The second main problem that stems from the constructivist's strategy to rehabilitate the ideational is the atomistic, reifying approach to the ideational it has yielded. Adding one more explanatory factor certainly opened up the list beyond power and interests, but it has also turned it into an endlessly expanding list onto which new items are constantly added, depending on where one is schooled—norms, ideas, rules, knowledge, cognition, beliefs, *principled* beliefs, attitudes, values, ideology, culture, symbols—all of which point to the importance of the ideational, to be sure, but there comes a point where it becomes difficult to discern between them and to know which does what, or, more importantly, whether we are all still talking about the same thing. The more fundamental problem that stems from this one-more-on-the-list-of-causes approach has been a tendency to treat "ideas" (used here as a shorthand for all of the above) as things, almost as positive entities. And I do not mean only that, in wanting to salvage their causal purchase, they have been reduced to mechanistic forces that effect change, like a cue pushes the ball around the billiard board, but that it is as though these ideational factors themselves look like as many dispersed, atomized entities. Both that which links them together, and that which constitutes them, namely, discourse, fades out of sight. As Karen Litfin (1994, 3) already pointed out in what was effectively the first systematic and extensive attempt to deploy a discursive approach in the study of international politics, these ideational factors are singularly disembedded, not so much from the social or institutional but from the *discursive* relations in which they first took shape.[11] Hence it is not that discourses serve to "communicate" ready-made ideas; it is that they constitute them (Schmidt 2008). Ideas do not exist outside of discourses, and discourses are not merely their "containers" or "transmitters." Consequently, ideas have become reified. That is, they are treated as neatly demarcated, finished products that are used to explain other things nonideational (such as differences in labor practices in the textile industry; Biernecki 1995), but the process of their own making is rapidly eclipsed. Having began to deconstruct our social "givens," it is as though the "given" has been merely shifted—away from "interests" and even "norms," for sure, but toward "ideas" and "identities" instead (see Zehfuss 2001 or Hansen 2006 for the latter critique). In other words, the stuff of it all—namely, discourse—is rarely taken as the analytical material itself, into which to sink one's scholarly teeth. This is what I propose to do in this book. Most importantly, the making of meaning is far messier, and it rarely yields such clear-cut, fixed things. The focus on meaning restores the dynamic and bloody processes that constituted these ideas in the first place. For ideas are fragile constructs; even when they endure, they are but temporary fixations of signifiers, always potentially undone by possible rearticulations.

The Denaturalizing Task of Critique

The discursive understanding of power proposed in this book takes conflict rather than cooperation to be the main modality of political life, and ruptures rather than continuities as the determining moments of history. If the social order is not pregiven and constantly being reasserted, politics is not simply an additional layer tagged onto an inherent order wherein preexisting social conflicts would be managed or resolved. Politics is rather the struggle to shape this order itself. In placing the focus on the articulatory struggles underlying the making of meaning and social formations, discourse theory epistemologically foregrounds the ontological "primacy of politics" (Howarth 2000, 104). Consequently, in terms of research design, conflicts are especially useful as catalysts for exposing particular constellations of meanings and power. Thus the interesting moment in a discourse perspective is not when cooperation runs smoothly but rather *when it breaks down*—hence the salience of the whaling issue, where the lines of the battle are drawn out in the open. The whaling regime, which is increasingly polarized between an anti- and a pro-whaling faction, presents a case of failed cooperation. Whaling is banned internationally, yet it is actually increasing on the ground. The failure of cooperation is rooted in a struggle to impose what international whaling relations should really be about: saving whales or managing whaling. Thus a discourse perspective, far from taking cooperation as its starting point, begins by questioning it. More broadly, the fact of cooperation raises the question of what power differentials needed to be smoothed over in order to achieve it.

It will be clear by now by now that the discursive approach deployed in this book belongs to a Foucauldian rather than the Habermassian lineage.[12] The attention to discourse developed in the wake of Jurgen Habermas's theory of communicative action are geared to the possibility of arriving at a reasoned consensus (Risse 2006, Kratochwil 1989, Onuf 1989). This possibility, in turn, is grounded in the Kantian belief in reason as a constitutive feature of the human essence, which grows old roots in Western philosophy reaching back, beyond Plato, to the Pre-Socratic thinkers. When the question of essences is ontologically and epistemologically suspended, as it is in this discursive approach, the ground wears thin under the possibility of building consensus. Therein lies the crux of the difference in the way of envisaging norms that has prevailed in the constructivist literature, where they tend to be seen as "neutral" or generally even "beneficial" (such as human rights or environmental norms) rather than as effects of power.

Taking discourses as an object of study aims to denaturalize what we assume to be right, or, to put it in yet another way, to dissolve the *doxa* we unquestioningly dwell in (Wacquant 2001). It is to raise the question of how the categories of a discourse (the objects and subjects it produces) are wielded in the production of "objectivity" and "truth." Hence faced with a discourse, the task is not to query whether its statements are true but to study how its "truths" are mobilized and meted out.[13] The question I ask in this book is not which discourse, whaling or anti-whaling, is the more truthful. Rather, I consider their truth *effects*. For "the truth" is potent; its power is wielded in particular discursive economies of power. Thus it becomes necessary to assert the relativity of truth claims and to consider them in relation to the particular configuration of power relations within which they obtain. More generally, studying discourses is a means to taking a critical step out of what the discourses actually say, in order to observe what they do.

Problematizing the Subject

The Duality of Discourses

A social actor is also a *speaking* actor, and therefore the subject—the I/we—of particular discourses. In speaking, the actor does things; it achieves certain concrete, practical results. However, it also positions itself in relation to other speaking actors, it marks itself in a particular way—that is, it also "does" something for its identity. Hence at these two levels of action and identity, speaking, making sense, is a modality of the actor's agency. Discourses are *enabling*, in that they allow the actors to act in the social world. They are by the same token *constraining* (Giddens 1979, 1990, 1991). In order to make sense to others, the social actor has to both speak and behave according to shared social conventions. It has to observe the rules of this syntax, so that its statements and actions may be deciphered by these social others.

Discourses are thus constraining in terms of the technical requirements of sense making, that is, in terms of "being understood." They are constraining also in a more moral, normative sense. For discourses are the repository for the benchmarks of good and bad behavior; they contain a society's values and norms—that is, its modalities of social regulation. For norms are, in Foucault's words (2003, 38), the "natural rule," through which society is regularized, and discourses are its structures of normalization. The normal and the deviant are located within hegemonic discourses (Foucault 1967). Whaling was normal until the mid 1960s. The new anti-whaling discourse displaced the norm, such that it became "unacceptable," even "barbaric." Normality is thus relative and discursively ordained. Most importantly here, where the norm lies is an effect of power.[14]

Subject-Positions versus Subjectivities

Approaching social actors as speaking actors has significant repercussions for the study of identity, a defining concern for the constructivist scholarship (Price and Reus-Smit 1998, Wendt 1999, Guzzini 2000, Wight 2004, Flockhart 2006). It shifts the focus away from the production of "subjectivities" to "subject-positions." Against the evacuation of agency that had tended to occur in the Marxian appraisal of the subject as produced exclusively by material/social structures, and subsequently in Foucault's own approach to discourse, discourse theory distinguishes between "subject-positions" and "political subjectivities." Only subject-positions are produced by discourses; social actors' political *subjectivities*, on the other hand, cannot be reduced to discursive production (Howarth 1995, 123). This distinction is key to opening up the space for a relational approach to international relations. In other words, a *subject-position* refers to a *position within a discourse*. By contrast, "subjectivity" is a much more extensive and, consequently, a more unwieldy concept, one that can include things that elude processes of symbolization (such as bodily functions, or Lacan's categories of "the real" and "the imaginary"). Both similarly point to the making of identities. What I attempt to show in this book is that, in separating out these two dimensions, the discursive approach steers clear of many of the hurdles that have encumbered the study of identity in international politics. Once that distinction between subjectivity and subject-positions has been drawn, it becomes possible to approach "every subject-position [as] a discursive position" (Laclau and Mouffe 1985, 115), and as a discursive position *only*. That is, it becomes possible to bracket issues of subjectivity.

To return to our social-speaking actor, how then is its identity discursively produced? I argue that in *stepping into* a particular subject-position carved out by a discourse, in taking on the "I/we" of that discourse, the actor's identity is produced in a very specific way. In doing so, the subject is establishing itself as the subject speaking the particular discourse, such as the anti-whaling discourse, and thereby marking itself as an anti-whaler. This is very different from internalizing the norm that condemns whaling. For a start, it is a much more active process. The actor is making its own identity rather than receiving a norm from an external authoritative "socializing agent" (Checkel 2005, 813). Moreover, it no longer becomes necessary to assess "how much" the socialized agent "truly has" internalized the norm and been "really" persuaded, as opposed to behaving in that way because of the instrumental payoffs tied to it, something which may be at any rate very difficult to prove without opening up the actor's "head." What matters is, quite simply, what the actor says. If it speaks the anti-whaling discourse, then it is marking itself as an anti-whaling subject. From there, the analysis can begin to examine what made this possible in the first place, and what exactly the actor is accomplishing in positioning itself thus. Most importantly, focusing on subject-positions gets much closer to explaining how actors' identities and interests are actually constituted. For the norms actors adopt do not effectively define them, even while they may be fully internalized, and if they conform their behavior to it. What *actually* defines, what shows that they recognize themselves in that norm, are the discourses they speak. The discourses they speak mark who they are, both to themselves and to social others.

This significantly opens up the analytical scope. First, if the social actor is a speaking actor, and a social system is one where discourses circulate, then the international system can be approached as a social system. Of course, approaching the "society of states" as a space of social interactions is nothing new, growing deep roots, beyond constructivism, in the English school, where the expression was first coined (Bull 1977). Both of these, however, largely overlook the *social* dynamics, because they tend to reify actors' identities—in constructivism, because of the treatment of identity as "given" rather than as a dynamic process of *identification* (see Zehfuss 2001 for this critique); in the English school, because a largely state-centric focus has barred from the analysis non-state actors who may, as the whaling issue will show, quite literally step into states' subject-positions in international organizations (see chapter 8). Social dynamics necessarily become much more central to the analysis when they are seen as actually producing actors' identities.

Consequently, states are seen to be part of a dense social fabric, where the way individuals interact in everyday life can shape the course of interstate relations. By placing at the center of its focus the very medium through which all interactions occur, a discourse perspective properly foregrounds the relational dimension of international *relations*.

Second, the distinction between "subjectivity" and "subject-positions" resolves the levels-of-analysis problem in international relations. That social actors are speaking actors applies equally to individuals and corporate actors, to nongovernmental organizations (NGOs) as well as states. A significant advantage of a discourse perspective is that it enables the enquiry to travel the full length of the levels-of-analysis spectrum, from the individual to the state level. Individuals, NGOs, states, are all potentially the subjects (the I/we) of a discourse. Hence by approaching the subject as a discursive category, discourse theory introduces a degree of *equivalence between subjects* pertaining to otherwise different levels of analysis. This yields a certain flexibility in approaching the subject, and the possibility of tapping into theories of formation of the subject wherever relevant to understanding the subject under discussion.

Chapter Outline

The chapters of this book fall into three parts. Part I examines a past whaling world where the dominant discourse was about killing whales. The production and reproduction of the mirror opposite dominant discourse about saving whales is the object of parts II and III, respectively. This book is thus built around a rupture, between a whaling "before" and an anti-whaling "after." Juxtaposing side by side two completely opposite discourses about the very same resource is a way of utilizing the "jarring effect" that can sometimes stop us in our tracks and make us reconsider what we normally leave unquestioned—here, our prevailing notions about whales and whaling. It serves, in other words, to undertake the denaturalizing task of critique.

Retrieving a long-forgotten but not-so-long-gone whaling order from within our own whaling past in part I serves three main purposes. First, it sets up a foil for reflecting upon where we stand today. Our own whaling past is used to create the critical distance from which to examine the current anti-whaling order. If appraising historical contingency is indeed one of the main drivers of the turn toward the examination of "the social construction of," then the pertinence of the whaling issue is that it shows not only that things *could* have been otherwise but that, in this case, they actually *were*. The constructivist concern is unencumbered with any speculative nostalgia for alternative perspectives that could have prevailed if the world were a better place that can sometimes burden attempts to uncover subaltern perspectives and subjugated knowledges, to borrow Donna Haraway's (1991) term (see, for example, Shiva 1998).[15] In the case of whaling, we really were on a completely different path, one that was about hunting whales to extinction rather than saving them. Setting up this foil serves to draw out the historical relativity of the discourse that prevails today and the extent to which its "truths" are not absolutes but hold currency only in specific historical epochs.

The second important aim is to render visible a world that remains invisible by an effect of power. This "rendering visible" is key to the task of unmasking a particular form of power whose principle of effectiveness is its invisibility, namely, symbolic domination. Excavating the past whaling discourse draws into relief the functioning of the current anti-whaling discourse. For "our" past whaling world is not just "passed," it is actively forgotten. Amnesia constitutes a key mechanism of the anti-whaling discourse, whose main subject position, the anti-whaling "us," is constituted through the active denial of any similarity with the whaling "them." A third methodological aim in revealing this whaling world is to draw out the unity of a social system within which discursive and material practices constitute each other. This shows how an individual is connected to broader structures of normalization that are deployed from the individual to the global level. Indeed, this whaling discourse is what holds the whole social system together. It is also what links the various levels of analysis. Part I thus unpacks the whaling order at three successive levels of analysis, proceeding from the individual to the global level. The following chapter begins on the ground, as it were. There examination of concrete whaling practices brings to light the extent to which whale parts were pervasive in the life of the modern individual. Chapter 2 thus begins by surveying the various forms of consumption of the whale, thereby revealing both the omnipresence of the whale resource in everyday lives and the varieties of whaling around the world, some of which continue today.

Moreover, chapter 2 addresses the question of material interests by analyzing the political economy of modern whaling. Discourse is not center stage in this chapter. It examines, first, the way in which the industry modernized its productive structures, in line with rapidly industrializing economies, and second, the factors that continued to fuel the demand for whale produce, thereby upholding the industry beyond the point where whaling itself had become unsustainable, both economically and ecologically. A key factor that accounts for the endurance of the modern whaling industry, despite plummeting whale stocks and technological innovations that progressively substituted the main uses for whale parts, was the phenomenon of "whaling nationalism," which saw more and more countries competing for fewer and fewer whales. In other words, far from turning away from whaling as the whales declined, countries wanted to whale *even more*. Methodologically, by showing that the interest in whaling endured beyond the point where it was sustained by the configuration of economic interests, chapter 2 disjoins the *material* whaling interests from the *interest in* whaling, thereby clearing the ground for a discursive approach. In sum, the West remained interested in whaling well after Western commercial interest had abandoned the trade, because its interest was framed by an entrenched whaling discourse.

Chapter 3 continues to unravel the significance of whaling to this past whaling world. It examines the nexus of whaling and the state, so as to tease out the role of whaling in processes that shaped the modern nation-state. It deploys Foucault's distinction between "sovereignty" and "governmentality" as two different lenses for drawing out the connections between whaling and state practices. The chapter then analyzes the particular constellation of power/knowledge in which whaling was enmeshed. There it examines the relationships between the state, science, and the whaling industry that led to the development of cetology (the science of whales) within national structures of knowledge and power centered around the whale.

Chapter 4 analyzes the emergence of a common international whaling discourse that took shape around the development of whaling regulations. Starting from measures in place at the national level, it retraces the halting progression toward international whaling regulations that eventually yielded, in 1946, the IWC. Methodologically, the chapter takes issue with both regime theory's traditional accounts of the international politics of whaling, in both its neorealist and neoliberal institutionalist versions, and shows the need for a properly sociological account of the dynamics of interstate actions, *even when* they are driven mainly by competition rather than cooperation, which are both inherently social phenomena. It then deploys such an approach, using Pierre Bourdieu's concept of *field* and his analysis of interests. Stretching interests toward a sociological understanding puts further pressure on the narrow acception of the concept which has prevailed in the study of international politics. In the subsequent analysis of the "society of whaling states," states are envisaged as *social-speaking actors*, whose identity is constituted not only by the way they interact with one another within a social field but also by their *belonging* to the particular society it defines. The chapter shows that the dynamics of state belonging are a key explanatory factor in accounting for the ways in which states have behaved within this international organization. Both when it was the so-called club of whalers and subsequently when it became a club of *anti*-whalers, states have been consistently driven by their wanting to belong to it. For many states, being an IWC member is not about material payoffs, nor about making international cooperation work, nor even about what happens to the whales. Rather, what matters is where the *nomos* lies and to be seen to be embracing it by speaking the discourse that encapsulates it—then the whaling discourse, now the anti-whaling discourse. This explains how states have been able to so easily switch their expectations as to what the organization does (regulating whaling or precluding it) and take on radically opposite discourses about whaling. The IWC is a stage for performing states' socially constituted identities. Then, the "right" sort of state to be was a whaling state, and now, it is an anti-whaling state, a point which will be further developed in chapter 10.

Part II then appraises the birth of a mirror opposite, anti-whaling order. Chapter 5 analyzes the production of a new, dominant global discourse about whales that featured them as endangered, intelligent, and extraordinary mammals that needed to be saved from the whalers. The main theoretical issue foregrounded in this chapter is how to address the question of normative change in international relations, which has been approached from two different fronts, namely, the global environmental politics scholarship and the literature appraising the role of non-state actors. The chapter thus begins by positioning the analysis in relation to these literatures, as well as the field of critical security studies, which has so far been the most attuned to the role of discourse in international politics. The anti-whaling discourse challenges the latter's selection of dominant discourses, which tend to be reduced to state discourses. In this case, the national security discourse was a *whaling* discourse. The anti-whaling discourse, by contrast, arose from the margins of political life and successfully imposed itself upon a state from the ground up. The brunt of my critique of the literatures on normative change is that an excessive agentcentrism has reduced their ability to appraise the productive, structural power of discourse, which is more than a weapon in political actors' arsenal: it actually produces their social identities. The chapter introduces some conceptual tools to analyze the anti-whaling discourse, notably *articulation* and *interpellation*, and, from the discursive policy literature, *story-lines* and *discourse coalitions*.

The analysis of the discourse unfolds in two parts. The first part shows how the anti-whaling discourse rose to prevalence because it welded together two, preexisting metanarratives: that of the Cold War discourse on capitalism and democracy and that of a nascent environmental discourse. Specifically, the chapter analyzes the double synecdochic move that fixed the whale signifier in this new discourse in such a way that saving the whales became shorthand for saving endangered species, and the endangered planet as a whole. The second part of the chapter examines what the new discourse actually achieved. First, at a specific juncture in the early 1970s in American politics, it created a vast discourse coalition of anti-whaling state and nonstate actors that extended to the international level. Once again, this illustrates the methodological importance of finding ways to cut across the domestic/international divide in order to encompass the various levels of analysis at play (Walker 1993). Second, it provided a specific script for "doing something" about the environment, thereby yielding a new grammar for environmental activism at large.

Chapter 6 analyzes the role played by science in the whaling regime in order to appraise its power. Science is approached as the authoritative discourse on truth, regulating both what can and should be known within particular discursive orders, or "regimes of truth" (Foucault 1980, 131). Given the failure of the community of whaling scientists to build a scientific consensus as the basis for policy making in the IWC, the chapter starts by questioning the epistemic community approach, which is grounded in assumptions about science's ability to attain "the truth" about an issue and, from there, to drive policy making disinterestedly forwards. It then examines how much science was able to weigh into the policy decisions about whaling management, in each of the three phases that saw the consolidation of a science of whale management. What the history of IWC science shows is that, despite significant improvements in whale science over these three phases, to the extent that it became a model for fisheries management elsewhere, it was increasingly *ignored* by policy makers. The chapter thus shows that the power of science to influence policy makers is actually relatively limited. It is constrained by the particular epoch or *episteme* within which both the science and the policies are produced. What it does not appear to have is the power to make policy makers step out of its underlying normative order or *nomos*.

The anti-whaling campaign, the object of chapter 7, was one of the first successful global environmental campaigns. That chapter examines in detail the series of actions and strategies with which activists succeeded in shifting the *nomos* underlying whale related discourses and practices. Anti-whaling NGOs won over to their cause, first, members of an increasingly wider and more global public, and second, many formerly whaling states, which passed legislation to protect whales at home and actively pursued anti-whaling policies internationally. NGOs achieved this by calling both individuals and states into the newly created anti-whaling subject-positions. A key strategy to denormalize whaling consisted in rewriting the discursive categories underpinning both the perceptions and practices of whaling. In this way the anti-whaling discourse successfully displaced the boundaries of the acceptable/unacceptable and even the legal/illegal. It also defined the categories through which whaling would be thereafter managed at the IWC, including the category that served to suspend it, namely, "commercial" whaling, distinguished from "aboriginal" whaling. The chapter ends on the strategies deployed to engineer the 1982 commercial whaling moratorium vote at the IWC, which marked the final stage of the transformation of the "society of whaling states" into a "society of anti-whaling states." Methodologically, this account of the anti-whaling campaign is useful for expounding the difference between *actors* (individuals or states) and discursive *subjects* (or I/we).

Having examined different facets of the *production* of the dominant anti-whaling discourse, part III considers the factors sustaining its *reproduction*. A discursive approach entails simultaneously a series of theoretical commitments (discourse theory) and a method of analysis (discourse analysis). Chapter 8 undertakes an applied analysis of the anti-whaling discourse. Taking as its basis a boycott advertisement authored by a coalition of anti-whaling NGOs in 1974, and drawing parallels with a contemporary anti-whaling caricature, it examines the space of relations staked out by the anti-whaling discourse. The whales-object cast as passive victims conjures up two subjects set up in a binary, "them-whalers" and "us-anti-whalers." Thus, whereas the previous chapter showed how the new subject-position created by the anti-whaling discourse was mobilized in the campaign itself, this chapter examines more closely how it was actually carved out. This subject-position was tailored for the contemporary global individual consumer who cares about the environment. My main argument in this chapter is that the anti-whaling discourse has been able to last because it created identity categories that tapped into, and reinforced, existing representations that obtained in particular political economic relations—for example, between the United States and Japan. Schematically, the rise of Japan as an economic threat thus coincides with the representation of the Japanese as a threat to the whales. Methodologically, in terms of the discursive approach to the making of identity, the chapter draws out a key distinction between *subject*, that is, a space within a discourse, and *subjectivity*, which refers to the actual identity taken on by flesh-and-bone individuals.

The analysis in chapter 9 remains at the level of the individual and considers the consumptive practices that sustain the anti-whaling discourse to this day. It begins with a typology of "nonconsumptive uses of the whale" to use the discourse's own terminology, which mirrors the survey of consumptive uses underpinning the whaling order that opened chapter 2. The anti-whaler, this chapter will show, is constituted as much by what she or he says as by what she or he consumes. Particular image-based and virtual forms of consumption thus comprise another key factor enabling the reproduction of the anti-whaling discourse.

The last two chapters shift to the other end of the level-of-analysis spectrum and analyze the conflict between anti-whaling and pro-whaling states currently being played out at the IWC. This conflict is used as a catalyst to expose the dynamics of the confrontation between a dominant anti-whaling discourse and a pro-whaling discourse of resistance. Chapter 10 examines the ways in which states take on the anti-whaling discourse in order to position themselves within a broader society of states. Two key methodological aims are achieved in this chapter. First, the ways in which a particular discourse relates to broader discourses are explored. Second, and crucially for a discursive approach to the construction of state identities, the chapter elaborates the distinction between state *subject-positions* and state *identities*. In analyzing how states step into the anti-whaling subject-position, I hope to show how "subject-position" offers a far more adequate conceptual tool for the analysis of state identities than the concept of "identity," which is a concept first deployed at, and better left for, the individual level of analysis. "Subject-position" provides a way of analyzing state identities that excludes the dimension of subjectivity.

Chapter 11 serves as the counterpart to chapter 10. It examines the recent formation of a pro-whaling discourse developed *against* the anti-whaling discourse, as an "anti-anti-whaling" discourse. The subject-position carved out here is not simply the same old subject-position carried over from the past whaling order; rather it is one developed from a position of resistance. As a result, the whaling identity itself has been transformed. In other words, the subject-position proposed in this discourse is not simply a whaler's; it is a *pro*-whaler's. The analysis travels once again along all the levels of analysis, examining how the discourse has taken shape at the national level, within relations between states, in relations between states and substate communities, and lastly between substate whaling communities. At stake for this new, pro-whaling discourse is recovering the power to define their whaling identities.

I

The Whaling Order

The more I dive into this matter of whaling, and push my researches up to the very spring-head of it, so much the more am I impressed with its greatness and antiquity; and especially when I find so many great demi-gods and heroes, prophets of all sorts, who one way or another have shed distinction upon it....

But even stripped of these supernatural surmisings, there was enough in the earthly make and incontestable character of the monster to strike the imagination with unwonted power.

Herman Melville, *Moby-Dick*
(1851, chapters 82 and 41)

2

An International Political Economy of Modern Whaling

Whaling is an ancient craft that has been traced back to 15,000 B.C. (Stoett 1997). Yet, in a sense, *modern* whaling was born against all odds: after the discovery of mineral oil in Pennsylvania in 1859, whale oil was progressively replaced in its most important uses, lighting and lubrication. Despite this ostensible substitution, the next hundred years saw an unprecedented expansion of whaling and the emergence of the large-scale, industrial whaling properly known as the "modern whaling industry" (Tonnessen and Johnsen 1982), able to hold its place among the staples of the industrialization process well into the 1960s. How can the endurance of whaling in modern societies be explained? Before answering this question, the chapter begins by mapping out the various forms of whaling, according to the different ways of utilizing the parts of the whale, so as to introduce the reader to practices that have largely fallen into oblivion today. This typology of whale-related consumptive practices serves to illustrate the extent to which whale parts pervaded everyday life. Whaling, it will be shown, was deeply embedded in the productive and consumptive structures of modernizing societies. The chapter then analyzes the factors accounting for the whaling industry's persisting competitiveness and its ability to continue to attract capital until the 1960s, in spite of the ongoing threat of substitution by other raw materials. Two lines of explanation are pursued successively. The whaling industry remained competitive because it adapted its production practices, which are appraised in the second part of the chapter. Furthermore, it was able to continue to sell. The third part of the chapter thus considers the market for whale products and how these fared in relation to potential substitutes. Price fluctuations, however, only partially capture the continued interest in whaling, which was steeped in political and strategic considerations that defeat strict market logic. Hence the fourth and final part of the chapter considers the political and cultural significance of whaling as revealed by the wars of the twentieth century and in the more long-term phenomena of "whaling nationalisms."

In addressing the issue of material interests, this chapter operates a series of ground-clearing moves for the discursive approach deployed in the rest of the book. A widespread assumption is that the West stopped whaling simply because it had become uneconomical (Schneider and Pearce 2004, Mazzanti 2001, Stoett 1997, Davis and Gallman 1993, M'Gonigle 1980, Friends of the Earth [FoE] 1978, Scarff 1977, Small 1971).[1] In other words, material interests fully account for the West's withdrawal from whaling, such that the anti-whaling discourse would be nothing but a by-product of a reconfiguration of these interests—and therefore not worthy of study in its own right. What a closer examination of the international political economy of whaling draws out is, first, that the question is not why did whaling stop but rather why did it *last* so long, beyond the moment when it had become uneconomical. Second, it shows that the West remained interested in whaling even after it had largely pulled out. In other words, the end of whaling in the West was *not* coextensive with the end of the Western interest in whaling, and this interest is not reducible to its material interests. In the foregone whaling order, whaling was not only normal but it remained desirable, and Western countries were careful to cultivate the possibility of their resuming whaling at a later date. It would take another dominant discourse, the anti-whaling discourse, to rule out that possibility altogether by denormalizing whaling.

From Head to Tail: Using the Whale

The extent to which whale parts were ubiquitous in everyday life in the first half of the twentieth century is difficult to fathom today in the light of an extinct whaling industry. One needs to conjure up images of pipes, piano keys, cigarette holders, earrings, brooches, lipsticks, creams, candles, soaps, perfumes ... The modern individual was constantly in contact with whale-derived goods. Some languages still hold traces of this pervasiveness if we listen closely enough. In French the baleens of the whale—the keratinous plate hanging from the roof of the mouth of certain species—have given their name to the structures in umbrellas and women's corsets ("baleines"). The whaling industry was the equivalent of today's petroleum industry. The whale oil, extracted from the blubber, the bones, and the skin of most whales, comprised the fuel, and the baleens or "whalebones," as they were sometimes known, provided the equivalent of today's plastic—a robust and elastic manufacturing material.[2] Heated and shaped, they would make anything from corset stays, umbrella ribs, ramrods, and fishing rods to buggy whips and carriage springs. Cut into thin strips, they were used for sieves, nets, and brushes; further shredded, they made furniture upholstery. As for whale oil, it was used for lighting—the streets of London and New York were lit with whale oil until the late nineteenth century (Jackson 1978, 122–125; Creighton 1995, 14). As lubricating oil, it served in manufacture of soaps and cosmetics, as well as varnishes and paints. It was also employed in the tanning and textile industries and, notably, was important to the soaring British textile industry throughout nineteenth century for the cleansing of wool (Jackson 1978, 120). Furthermore, the wax-like substance in the head cavity of the sperm whale yielded another valuable substance, known as "sperm oil" or "spermaceti." It procured both a high quality lubricant and a wax that, unlike whale oil, does not smoke when burned.[3] The technique for separating out spermaceti was mastered in the eighteenth century, but its lubricating qualities were only fully exploited in the late twentieth century, as we shall see. Thus while whale oil illuminated city streets, spermaceti candles glowed in the drawing rooms. The ambergris, produced by the sperm whale's intestines, was used in perfumery and, briefly, in cookery (as flavoring fat). Other byproducts of the modern whaling industry included whale meat, sporadically used as dog and fur farm food, fishing bait, and cattle meal, and the bones, which made fertilizers.

The ways of utilizing whales have been as varied and ingenious as the forms of whaling around the world. *Modern* whaling—the whaling that almost drove whales to extinction—can be seen on a line of continuum that begins with the *opportunistic* whaling of, for example, the eighth-century Norsemen, who first developed longer handheld spears better adapted to a larger prey, extending to the *traditional* whaling, first associated with the Basques in the Bay of Biscay in the eleventh to twelfth centuries, who organized hunting expeditions that exclusively targeted whales. These three forms of whaling are not to be understood as fixed categories, nor as occurring during neatly carved out historical periods, but rather as schematic ways of utilizing the whale that may coexist at one time. Both opportunistic and traditional whaling still occur today. It is not rare for incidents of whale use to crop up in the local news: a whale stranded in Bangladesh a few years ago, for example, was sliced up and shared among the local villagers (Associated Press 2002). As for traditional whaling, it continues today in many coastal communities that have always turned to the sea for their resources. In Japan, four coastal towns still practice traditional whaling (Institute for Cetacean Research [ICR] 1996a). There, anthropologists have traced back 800 years of dietary customs involving ritual celebrations of the animal's soul together with consumption of the meat (ICR 1988a, 1988c; Kalland 1989; Hiraguchi 2003). Larger whales were hunted as of 1675 after the discovery of new methods in the coastal town of Taiji. On the Island of Bequia, in St. Vincent and the Grenadines, the pilot whale, known as the "blackfish," is considered a bounty. In fact, the only form of whaling that has mostly disappeared is the large-scale modern factory whaling.

Geographically, whaling evolved increasingly further from the coasts. Thus the era of bay fisheries whaling, where the whales were caught along the shores, was succeeded by seaborne whaling, followed by an the era of *pelagic* (high seas) whaling, where the development of factory ships reduced the need to return to the land to unload and refuel and thus allowed for the exploitation of the remote waters of Antarctica. The whaling trade experienced a Basque, Dutch, British, American, Norwegian, and Japanese predominance, overlapping to some extent, with experience passing on from one group of whalers to the next. Every episode brought new people, new modes of organization, and sometimes new methods. Each one altered the geography of whaling: thus the Arctic—"Northern"—fishery, was followed by the Southern fishery, the Pacific fishery, the North Atlantic fishery. The Antarctic fishery marked the final chapter in the history of modern whaling, which is the story of a self-destructing trade. For modern whaling arose from the repetition of the same tragic cycle where the whalers found the whales, hunted them down, and moved on, until they reached Antarctica, the whales' most important breeding and feeding waters. Soon there would no longer be enough whales to sustain the trade.

A key difference between traditional and modern whaling revolves around their ways of utilizing the animal. In traditional whaling, the whale tends to be harvested in its entirety. It is, for one, an important food resource. Besides the whale meat, consumed in all whaling communities, the nutritious layer of skin and blubber (the fat under the skin)—or "mattak" as the Inuit know it—is enjoyed around the Arctic

(Greenland and Iceland) and in Japan, but not in Norway. The baleen provides construction material for the people of the Arctic, and the skin procures leather to the Alaskan Inuit (Freeman 2000). Both the bones and ivory of toothed whales are used as carving material by "scrimshaw" artists in the Northern Arctic and the South Pacific. Modern whaling, by contrast, made a very targeted, limited, and consequently wasteful use of the whale: it sought mainly the oils, and the baleens to a lesser extent. This meant that out of a blue whale weighing between 90 and 150 tons, 9 to 10 tons of whale oil would be extracted; the rest would be dumped back at sea. Another difference is that, whereas traditional whaling tended to be a localized practice, deeply embedded in the particular structures of meaning and social relations of small coastal communities, modern whaling increasingly evolved into an activity organized at the national level and brought under the gaze of the state, as we shall see in the next two chapters. This has sometimes been a source of social conflict at the local level. In the Finnmark region of Northern Norway, considered the birthplace of modern whaling, whalers were perceived by a poor, fishing proletariat as wealthy capitalists (i.e., boat owners) from the South who were coming to take over their fish (Tonnessen and Johnsen 1982, 63–66).[4] The fishermen accused the whalers of having driven the capelin from the seas. Years of tense relationships and even rioting eventually led to the proclamation of a whaling ban off the coast of Finnmark in 1902.

The rise of modern whaling was driven not by the needs of human consumption but by industrial demand. Its main products were the oils and baleens, two raw materials that went to the heart of industrialization. One exception is Japan, where the meat has always been a key motivation for whaling. Even in Norway, where the main produce of whaling today is whale meat, the formidable expansion of the whaling industry in the early twentieth century was mainly spurred by the price of whale oil. For this reason whaling historians Tonnessen and Johnsen (1982) distinguish between an inherently wasteful Western whaling, which was essentially for industrial purposes, and Japanese whaling, where human consumption remained a key driver throughout. Japan thus blurs the distinctions between "traditional" and "modern" whaling, in that the factory ships—the infamous emblem of modern whaling in the West today—that are still being operated for "scientific" whaling also yield whale meat. (Norwegian whaling, by contrast, no longer uses factory ships but rather smaller whalers that are comparable to the vessels employed in the four Japanese coastal communities.) Thus, in Japan, factory whaling coexists today alongside traditional whaling.

Production Structures

Substitution was a long, drawn-out menace for the whaling industry (Schneider and Pearce 2004). It began in fact prior to the discovery of mineral oil with, notably, the import of rapeseed oil, which began replacing whale oil in British textile manufacturing in 1821 (Jackson 1978). Yet whaling thrived throughout the late nineteenth century, because it was able to meet the soaring demands for raw materials wrought by the rise of mass consumption. Whaling adapted its ancient practices and evolved into a proper modern industry, such that the whaling industry soon became a marker of being a modern, industrialized country. Here I rapidly survey the milestones of this modernization process. The 1860s marked a turning point: steamboats were used for the first time, and new killing methods improved catch efficiency (notably the exploding grenade harpoon introduced by the Norwegian Svend Foyn in 1868). Large-scale factory ships were introduced in 1903, thus enabling the whale to be processed directly on board and eliminating the need to tow the massive animals back to the whaling station. These also meant longer and more efficient hunting periods. For once processed, the risk of putrefaction was eliminated, and the whale produce could be stored on board. The first Antarctic whaling station was established in South Georgia in 1904, inaugurating the era of Antarctic whaling. Whaling could now take place farther, for longer, and in waters where the whales were prolific. By 1913 there were six landing stations in the Antarctic and twenty-one floating factories, and by the 1920s, only a small proportion of whaling was still conducted close to shore (Donovan 1995). Factory whaling was further improved in 1925 by the Norwegian "stern slipway," a mobile lower deck that facilitated the hauling of the carcass on board. It allowed industrial processing to take place in any condition (and no longer merely in the calm waters previously required for lifting the animal onto the upper deck). The production of oil increased nearly tenfold with the stern slipway. The number of catches also almost trebled as well: the forty-one factory ships that set off in 1930 caught 43,129 whales, compared with 15,000 in 1914 (Scarff 1977, 347). As the whales became harder to catch as the stocks became depleted (as of the early 1930s), the processing techniques were improved so as to extract more oil from a single whale, such that volume increased more rapidly than the number of catches. Overall, these modernized methods yielded large production increases: on average, in one day of the 1927–28 whaling season, each factory ship produced 4,824 barrels of oil with 3.92 whales per day, increasing to 26,714 barrels with 10.45 whales per day in 1938–39 (Tonnessen and Johnsen 1982, 333).

The whaling industry remained competitive because of technological innovations that provided vital new outlets for whale produce. Technological innovation was in fact a double-edged sword for the whaling industry, for it effectively buoyed up an industry that was constantly threatened by cheaper substitutes, which had themselves been brought about by a series of technological innovations. In the 1850s, the German chemist Justus Von Liebig, considered the founder of modern agrochemistry, invented a method for producing meat extract from whale meat, which would be used to add a "meat flavor" to sauces (Tonnessen and Johnsen 1982, 724). This yielded the first "meat cubes" in 1882. Repatented, the product saw widespread use in the years 1958 to 1961, with the Nestle Company as its largest consumer. Alfred Nobel's invention of dynamite (nitroglycerin) in 1860 created a more ominous outlet for the whale oil's glycerol (fatty acids). Third, the process of hydrogenation discovered at the turn of the twentieth century meant that whale oil could be transformed into solid fat and thereby replace the use of tallow (cattle fat) in the manufacture of soap and margarine, two emblematic products of rising mass consumption. As for sperm oil, its future was secured well into the 1970s in defense and space technologies by the development of sulfurization techniques, which processed it into one of the best lubricants ever known, especially resistant to extremes of temperature and pressure. A final example of a modern use was the pharmaceutical industry's discovery of the whale's hypophisis (a gland beneath the rear brain), containing a hormone (adrenocorticotropic hormone) used in the treatment of rheumatic ailments (particularly arthritis). Pharmaceutical demand peaked as late as 1951 and 1957 (Tonnessen and Johnsen 1982, 723).

The whaling industry's global reach was another important competitive advantage. Whale oil was a global commodity, first, in that its distribution was global. An inherently mobile raw material, it could be rapidly freighted around the globe to the highest bidder as it was being produced. This was whale oil's edge over other oils, both mineral and vegetable, a large proportion of which tended to be consumed by the producer countries (Tonnessen and Johnsen 1982, 231). Furthermore, its seasonal nature meant that most of the production reached the market at the same time, that it could be bought all at once and stockpiled. This, in addition to its low price, is what enabled whale oil to play a larger role on the oil market than its actual volume warranted (Tonnessen and Johnsen 1982, 231). Second, production was organized on a global scale, both in terms of labor and capital. Whaling crews were typically international, a trait fictionalized in Melville's *Moby-Dick*. Norwegian gunners would sell their highly sought after skills to British, Dutch, Japanese, and then Soviet whaling captains. Norwegian whaling enterprises, in turn, were funded by British and German capital. Chilean and Argentinean whaling were for their part backed by Norwegian capital (Central Intelligence Agency [CIA] 1956, 33). This globalization of production continued even as the industry's main produce shifted from whale oil to whale meat, as we shall see. In the 1960s the Japanese whaling industry forged a set of transnational links with firms in other countries through purchase and investments. The large Japanese market for whale meat was supplied in the 1970s by Brazilian, Chilean, Icelandic, South Korean, Soviet, and Taiwanese, in addition to Japanese, firms (Peterson 1992).

An important attraction for investors was that the whaling industry had extremely low barriers to entry, as ships were converted into whaling vessels at a relatively low cost. The slightest increase in profits drew ships in from other trades, creating a flexibility unknown to land-based industries (where capital goods were more permanent). A ship merchant such as Aristotle Onassis, who was famously responsible for one of the most ferocious episodes of "pirate whaling" (which escaped all states' jurisdiction) from 1950 to 1954, could convert his ships into whaling factories while profits were high. Once these dropped, the same ships would be used for cargo transportation. This created significant instability in the trade, as rising prices had to be shared with newcomers out for a quick kill. Another attraction for investors was that, in whaling, profit returns were immediate, compared with the slower maturation of agricultural investments, notably in vegetable oils. As high whale oil prices signaled a period of expansion for the whaling industry, understanding how whaling was able to last well into the 1970s requires examining what sustained the peaks on the whale oil price curve (see figure 2.1).

Historically, the whaling industry was upheld by the formation of large industrial complexes characteristic of the second industrial revolu-

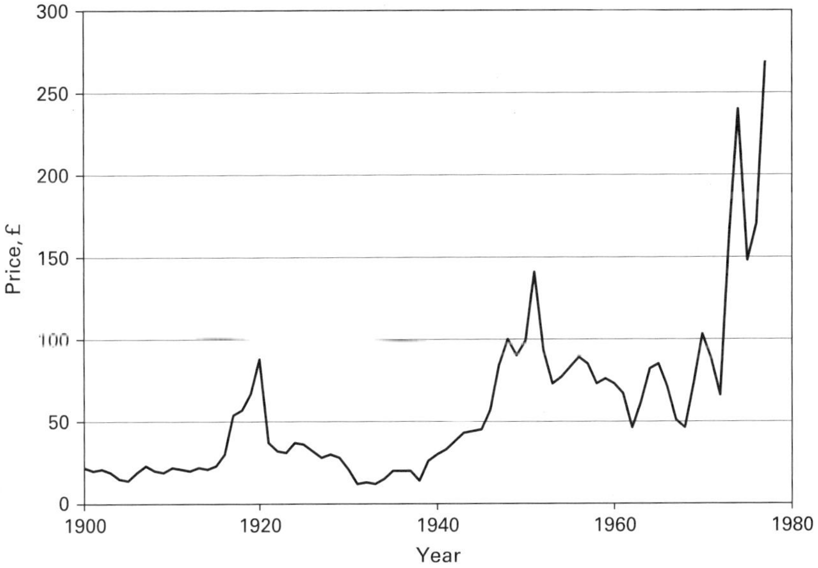

Figure 2.1
Prices of whale oil (in pounds sterling). Data source: Tonnessen and Johnsen (1982).

tion. This phase of industrialization saw the concentration of productive means along either a "vertical" or a "horizontal" axis, to draw on the language of economic historians. The horizontal merging of the soap and margarine industries in the early twentieth century directly impacted the whaling industry, as both were key clients. It enabled the producers of soap and margarine each to diversify their production (in a context when the demand for both products was still unstable) and together to consolidate the position of the soap and margarine industries as a buyer on the fats market. By 1920 there were four main groups left in the British soap and margarine industry—Crosfield, Gossage, Lever Brothers, and Watson. By 1927, these had been reduced to three, and on January 1, 1930, the gigantic Unilever Group was born (Tonnessen and Johnsen 1982, 235). A key concern was securing adequate supplies on a highly competitive market, and even before the birth of Unilever, the soap industry had sought to bring as much of the production process as possible in-house through vertical integration. The Lever Brothers, in particular, could not get enough whale oil for their soap production (Jackson 1978, 211). The first step was to buy a patent for hydrogenation, thereby reversing the dependency with the whaling industries, which had hitherto held a monopoly over the hydrogenation process. The large Norwegian whaling company De-No-Fa, for example, had been the main supplier of hardened whale fat to the Lever Brothers. The soap industry thus became an indispensable buyer for the whaling industry, and British and Norwegian whalers vied for the business of the Lever Brothers in the interwar period. To shore up against insecure supplies, the Lever Brothers stockpiled the oil. The integration was completed in 1919 when the Lever Brothers purchased the Southern Whaling Company to secure their own supplies. The bond was thus sealed between the whaling industry and a nascent large-scale agro-industry.

Demand for Whale Produce

The Role of Whale Oil on the Fats Market: 1900–1940s

The introduction of whale oil into the making of soaps and margarine warrants closer inspection, as it accounts for the persistence of a product that was otherwise destined "to disappear from the world market," according to whaling historians Tonnessen and Johnsen (1982, 229). For in the mid- to late nineteenth century, the need for fats was largely covered by cattle fats, both in their edible (butter, lard, milk) and non-edible (tallow) forms, as well as a few edible oils, particularly olive oil. The opening of the Western American prairie marked a shift from an outward- to an inward-oriented economic development and, consequently, the progressive withdrawal of U.S. capital from whaling. (Whaling did, however, continue on an ad hoc basis until it was outlawed by President Nixon in 1970.) Thus the United States, who had dominated world whaling for over a century, now became the largest producer and exporter of cattle fats. However, by the turn of the century, the fats supply was outpaced by the formidable expansion of the soap and margarine industry. Altogether, soaring demographics, a general rise in living standards, and a new concern with hygiene had generated a demand for soap on an unprecedented scale. Margarine, for its part, had been invented in France in the 1870s. It was composed essentially of animal fats (compound lard and tallow, as the solid fat was indispensable for its consistency), mixed with vegetable oils and milk. A cheap alternative to butter, it rapidly found its way into popular eating habits and soon became a staple item in working-class diets in Germany, Great Britain, Holland, and Scandinavia. Production took off in the early 1900s, and before World War I the manufacture of margarine exceeded that of butter in England and Scotland, for example (but not in Ireland; Jackson 1978, 184). By 1906, the combined demand for solid fats from booming soap and margarine industries created an exceptional fats shortage, driving the overall price of fatty raw materials to its highest peak for twenty years (Tonnessen and Johnsen 1982, 227–250). When tallow supplies proved unable to match industrial needs, the soap industry, in particular, turned to whale oil, the only oil whose price had remained stagnant since the late 1890s. Furthermore, it could now be processed into a solid fat using still very coarse hydrogenation technologies. It would take another two decades and considerable refinement of these processes before whale oil could be used in margarine; nonetheless, the impetus for their development had been provided. The year 1906 was a turning point for the whaling industry, as reflected in the price of whale oil (see figure 2.1). Fat hydrogenation, in turn, developed into a full-scale industry by 1911–1913, further stimulating the whaling industry: production more than doubled from 47,387 tons of whale oil in 1909–10 to 134,020 tons in 1913–14 (Jackson 1978, 178).

In 1929 scientists working both for the Lever Brothers and the Margarine Union discovered how to synthesize whale oil so that it could be used at 100 percent in the production of margarine. Significantly, these two firms merged the very next year to yield the huge agro-industrial corporation Unilever. At this point, margarine, rather than soap, became the most important outlet for whale oil. From then on, the decisive factors for the whaling industry would be the demand for margarine, the competition between margarine and butter, and the competition with vegetable oils, which could just as easily be hydrogenated to make margarine. Let us consider each factor in turn.

First, the rise of vegetable oils was an ongoing threat for the modern whaling industry. After rapeseed oil in the nineteenth century came cottonseed oil, launched on the fats market by Proctor and Gamble in 1911. In the 1930s, the danger came from soybeans. Once an expensive import from Manchuria, soybean oil production was being experimented with in the United States. So well did the crop fare on the American prairie that, over the 1930s, production increased twelvefold in the United States, such that, by 1938, the country had become an exporter of vegetable oils, thereby also collapsing the price of soybean oil (Tonnessen and Johnsen 1982, 457). Then came the tropical oils (coconut and palm) that were being developed in the colonies. The French, for one, were intent on sustaining the trade with their tropical colonies and soon discovered how to make a margarine entirely from palm oil (until then, it had not been possible to obtain the "buttery" texture without mixing in some measure of butter). Despite these pressures, whale oil held its place on the fats market: the record 1930–31 production yielded a volume of whale oil comparable with the total production of olive oil of France, Italy, and Spain combined (Tonnessen and Johnsen 1982, 370). The price of whale oil remained competitive throughout the 1930s compared with other edible oils (Tonnessen and Johnsen 1982, 742). All in all, the trade remained profitable: between 1924 and 1929, whale oil was sold at 32 pounds sterling a ton, while production costs were estimated between 15 and 18 pounds (Tonnessen and Johnsen 1982, 370).

As for the demand for margarine, second, it was so robust in the 1920s that production doubled from 520 thousand tons in 1913 to 1,050 thousand tons in 1927, thus securing an increased demand for whale oil. The whaling industry, however, offset these auspicious conditions by overproducing: production skyrocketed from 1 million barrels of whale oil in 1930 to 3.5 million in 1931. The year 1931 marked both a record and a turning point in the history of modern whaling: never again would the industry be able to produce as much oil, despite a constant increase in fleets and materiél (Scarff 1977, 350). It was also a tipping point for whale populations, which had been depleted beyond recovery by an industry that proved unable to respect the ecological limits of the resource upon which it depended. That same year, the prices of whale oil plummeted (see figure 2.1) as the effects of overproduction were compounded by the Great Depression. Furthermore, margarine production experienced a lull in the early 1930s due to the competition with butter. For the price of butter had been halved during the late 1920s under the pressures from cheap margarine. As a result, butter became competitive once again in the early 1930s. In the United Kingdom, for example, butter consumption increased by 47.5 percent between 1929 and 1934, while margarine consumption fell by 37.8 percent (Jackson 1978, 220). In the same time, governments, concerned to protect both their dairy farmers and their foreign exchange reserves, began to place restrictions or taxes on the manufacture of margarine—in Holland in 1932, in Germany, Denmark, Norway, Sweden, and Finland in 1933, and in Czechoslovakia and Italy in 1934. At that point, Unilever's European sales were a third lower than in 1929.

Third, considering the competition between margarine and butter, after 1934, the butter mountain melted, and the margarine trade revived. This was another turning point for whaling, which had become, with coconut oil, the main fat in margarine. Whale oil became a new, strong player on the market, its price exceeding 17 pounds sterling a ton in 1935 (see figure 2.1). Hence at the end of the Great Depression, the role of whale oil in the world fat market was reversed: while its price previously had risen or fallen in accordance with the dominant position of vegetable oils, whale oil had now become an important factor in setting world oil prices—so much so that the Food and Agriculture Organization's (FAO's) International Institute of Agriculture, looking back at the fats market in the prewar period, drew the following conclusion:

> The increasing use of marine oils, especially whale oil, in the foodstuff industry, particularly in the manufacture of margarine, led to the direct competition with the animal fats produced in agriculture. As the crisis developed, and purchasing power declined, the competition became increasingly severe. At last the production of whale oil became one of the most important factors in the unstable conditions and crises which affected the whole oil and fat market. Compared with production, the costs of which could be continuously reduced by means of improvement in technique and organisation, even the tropical plantations and the Manchurian Soya bean production were in a difficult position.

Thus in its analysis:

> Although the production of whale oil had a disastrous effect on the prices and markets of other oils and fats, the fact cannot be overlooked that, especially in Europe, fat supplies were greatly improved by whale oil production (quoted in Tonnessen and Johnsen 1982, 235).

The growth in margarine consumption (both in absolute terms and relative to butter) was sustained until the end of the decade, despite an improved economic climate, thereby buoying up whale oil prices. This was aided by periodic shortages on the fats market: the poor harvest of 1936, for example, increased the amount of vegetable oil consumed by producer countries, thereby relieving some of the competitive pressures for whale oil on the world oil market. Additional reprieves were brought by measures taken under the New Deal to curb agricultural production in the United States. For example, the area of land under cotton was limited under the 1933 Agricultural Adjustment Act, thereby decreasing the amount of cottonseed oil on the market. All in all, despite the competition, the place of whale oil on the fats market seemed relatively secure at the onset of the most destructive periods of the history of modern whaling known as the pelagic era.

From Whale Oil to Whale Meat: 1947–1970s

The last period of expansion for the whaling industry was triggered by yet another fats shortage, which hit war-torn economies in 1947 under the cumulated effects of a poor harvest, a soaring demand for raw materials at large in a world in reconstruction, and a shortage of vegetable oils. This was due, in turn, to the suspension of agriculture during the war, coupled with an increased consumption by producer countries. That year, the newly established FAO estimated an import deficit of about 55 percent for Europe alone and a worldwide shortfall in fat production of 3 million tons (Tonnessen and Johnson 1982, 528). In a context where domestic fats production could not keep up with the swelling pressures of a general rise in population and living standards, whale oil was in high demand, contributing 12 percent of fat imports in 1947. The price of whale oil would remain on a steep incline for a couple of years, almost tripling from 57 pounds sterling per ton in 1946 to 141 in 1951. These high prices, in turn, coupled with the disappearance of a large number of competitors, either because their fleets were destroyed or because they were barred from whaling (German whalers in particular), served as a strong incentive to set out whaling. Indeed, in the immediate aftermath of the war, everybody, it seemed, wanted to whale. A flurry of national whaling plans were announced, by the Netherlands, Japan, Germany, the United States, Argentina, Australia, the USSR, and Denmark, and even by countries that had never been involved in whaling, such as Sweden, Italy, or even landlocked Austria (Tonnessen and Johnsen 1982, 522).

Many of these national whaling plans would never leave the shelf, for, in reality, there were no longer enough whales left in the oceans for a large-scale whaling industry. By the time the IWC finally introduced national quotas to limit the number of whales caught per country in 1962, so depleted were the whale stocks that these could not even be filled: Norway, who had dominated world whaling between the wars, could catch merely a third of its authorized catches in 1962, the United Kingdom filled two thirds of its quota, and Holland, only half (Tonnessen and Johnsen 1982, 611). The whaling industry had become overcapitalized, and the resource too scarce—there were simply too many whalers, too few whales. Key problems for the industry were the long life cycle and the low reproduction rates of whales, which created an inherent tension between the rates at which the whale stocks could replenish and the rate of expansion of the industry. Furthermore, although the price of whale oil did continue to rise throughout the 1960s and 1970s (see figure 2.1), once vegetable oils had recovered and become once again extremely competitive, combined with the spread of plastic, it became clear whale produce would soon be replaced in its main uses and demand would peter out. The British and Norwegian private interests thus withdrew from an uneconomical trade in 1964 and 1968, respectively, and the Dutch state stopped subsidizing its whaling industry in 1964.

Only Japan and the USSR succeeded in filling (or exceeding, in the latter's case) their national quotas in the 1961–62 and 1962–63 seasons. From then on, whaling largely shifted eastwards, as these two countries took over world whaling.[5] Whale oil continued to remain central to the military–industrial complex during the Cold War, as we shall see. These strategic considerations, coupled with the enduring prices of whale oil, help explain the continued interest of the USSR. As for Japan, its modern whaling industry obeyed a completely different economic logic altogether. For whale meat had always been a key product of Japanese whaling, in addition to the oil and baleens. It had consistently sustained its industrialization throughout the 1930s. As a result, the Japanese whaling industry was always less sensitive to the fluctuations of the whale oil market than Western whaling. It was also far less wasteful, because it used every part of the whale. Thus out of a small whale, such as the minke, which was of no interest to Western factory ships, Japanese whalers could produce five to eight tons of meat, in addition to the two to three tons of oil extracted from the blubber and bones. This centrality of whale meat to the Japanese whaling industry was further accentuated by the way in which Japan returned to whaling after World War II. In a devastated country whale meat was seen by General MacArthur as a means to alleviate severe food shortages. Whale meat was thus the prime motivation for setting occupied Japan back to whaling in 1946—so much so that, in the first stages of the "symbiotic relationship" (M'Gonigle 1980) between U.S. capital and Japanese whaling, the Japanese retained the meat, while the oil went to the U.S. companies in return for their financing the expeditions.

By the late 1940s, the production of whale meat for human consumption had become an almost exclusively Japanese monopoly, despite attempts to introduce whale meat in other countries. The huge quantities of wasted meat generated by prewar whaling now seemed indeed unacceptable to a hungry and rationed war-torn world. Upon resuming whaling, British whalers, seeing the success of whale meat in Japan, sought to launch it on the British markets in 1947, through a concerted effort with the government (Jackson 1978). Amid a minor blaze of publicity, Food Minister John Strachey waxed eloquent about the 600,000 tons of wasted quality meat, while a Dr. Edith Summerskill educated the British public to the virtues of whale meat. The Department of Scientific and Industrial Research and the Ministry of Food spent the three postwar seasons researching in the palatability of whale meat, which had been further improved by the introduction of the electric harpoon in the early 1950s. Whale meat, which was not subject to rationing, was served in hospitals and schools. Lyons Cornerhouses were serving 600 whale steaks a day in 1947 (FoE 1978, 57). However, despite favorable reception of canned "corned" whale meat at the British Food Fair, the British public never really took to whale meat. Thereafter it was used exclusively in the production of animal feed in the United Kingdom (such that, in 1978, the British firm Pet Foods Ltd. was still importing 2 percent of Japan's production of whale meat; FoE 1978). Although it never took hold in Western countries, the idea of a large-scale whale meat industry was still floating around in the 1970s, as evidenced by this piece of Canadian research whose conclusion would seem unfathomable in the West today: "It is quite possible that the average North American exposed to a properly prepared grade A cut of whale beef will welcome it as an excellent alternative to cows or steer beef" (quoted in Scarff 1977).

Whale meat sustained the formidable expansion of Japanese whaling through the collapse of whaling in the West. In fact, the ascent of Japanese whaling facilitated the West's exit from large-scale industrial whaling, as Japanese firms bought off Norwegian, British, and Dutch floating factories. It also decisively altered the production and export patterns of the whaling industry. As of 1961, meat production increased in inverse proportion to the production of oil. By 1967, the price of oil was 45 pounds sterling, that of meat, 125; and the total value of meat produced in the Antarctic was more than five times the total value of oil (Tonnessen and Johnsen 1982, 722). Whale meat also spurred Japanese whaling's global expansion: for example, the Brazilian fisheries firm COPESBRA was established in the mid-1970s in the Northern State of Paraiba as a subsidiary of Nippon Reizo Kabashiki Kaisha of Japan

(Stoett 1997, 75). And yet, although it was both the biggest whaling nation and the largest whale meat producer since the late 1950s, Japan could not seem to produce enough whale meat: in 1977, it was still importing 19.946 tons of whale meat, 85 percent originating from the USSR. Nor was Japan then the sole consumer of Soviet whale meat: Iceland, Brazil, Peru, South Africa, and Spain all counted among the importing countries for Soviet whale meat. Japanese whale oil, by contrast, was destined at 87 percent for export, notably to the Netherlands. That whale oil was considered a by-product of the meat production, and meat, rather than oil, was the key driver of Japanese whaling, is also illustrated by the lack of interest in sperm whales, whose meat is considered inedible in Japan. The renewed interest in sperm whale oil in the 1960s (see below) largely bypassed the Japanese whaling industry, for which sperm whale oil stagnated at 7.5 percent of its whale produce exports (destined mostly to the USSR).

Modern Whaling and the National Interest

Whaling and Wars

While the takeoff of modern whaling was triggered by the 1906 fats shortage, its good fortune was sealed by the First World War. More generally, wars tended to provide the most auspicious conditions for the industry's development throughout the twentieth century. The next chapter examines what this reveals about the connections between whaling and the making of the modern state. Here it is enough to consider that during the First World War whale oil first became a strategic raw material, implicated both directly and indirectly in the war effort. It was all at once a fuel, a key ingredient in the manufacture of explosives, and a raw material for basic industries that were placed under government protection, not least soap and margarine. Supplying whale oil thus became a matter of national security, and governments began stockpiling whale oil, an involvement in the whale oil market that would continue well beyond the war. Furthermore, controlling whale oil supplies became a key aspect of the German blockade in the Allies' maritime strategy. For when the war broke out, the Norwegians were producing 77 percent of all whale oil (against 16.7 percent produced in the British Empire) and selling over a third of their production to Germany (Tonnessen and Johnsen 1982, 292). The United Kingdom thus pressured a neutral Norway to grant it exclusive sales of whale oil, negotiations which were facilitated by the heavy involvement of British capital in the Norwegian whaling industry, notably in the hydrogenation plants.

Furthermore, the war's propitious effects lasted beyond the war itself. Wartime rationing and government concerns about food security had decisively contributed to the acceptance of whale oil as a staple ingredient of products of mass consumption. In soap, they swept aside the soap manufacturers' lingering misgivings about perceptions of "fishiness" and how these would impair their sales. From then on, whale oil became a staple ingredient in the soap manufacture. The war equally removed any qualms about using whale oil in margarine. At a time when agricultural production of butter was suspended and margarine imports disrupted, the production of cheap, locally based margarine became a priority for the British government. It was thus upon the request of a government concerned by the dependency on the import of margarine from Germany that the Lever Brothers turned to the production of margarine in 1914. The British margarine industry exploded, the consumption of margarine overtaking that of butter by 1916. Furthermore, the war created the conditions that would ensure continued interest in developing the synthesization technologies that were eventually perfected in 1929. For the German blockade had effectively created a situation of client monopoly that directly benefited the Lever Brothers as the main buyers of Norwegian whale oil. They accumulated large stocks of whale oil that then needed to be used up after the war, thus securing the place of whale oil in the production of soaps and eventually margarine.

Whale oil took on increasing strategic importance in the buildup to the Second World War. The CIA, in the 1956 report mentioned earlier, attempted to gauge the perceived importance of whale oil, and thus the significance of whaling, for the British government. The CIA estimated that it formed "a significant part of the U.K. food supply." It found that whale oil comprised 37 percent of the margarine content, 21 percent of the lard compound, and 13 percent of the soap content between 1932 and 1936, and these percentages "steadily increased" after 1936 (CIA 1956, 20). In 1938 the British government classified whale oil, alongside meat and sugar, as essential "national defence" commodities. Throughout the war itself, and although mineral oil had become the main fuel, whale oil derivatives remained "an important element in the manufacture of explosives and lubricant," according to the CIA. In 1938, Germany and the United Kingdom purchased 83 percent of the world whale output between them. Whale oil was once again at the heart of the war.

Its strategic control was so important that the British were prepared to concede rather liberal Norwegian exports of *fish* to Germany against rigid restrictions on whale oil. The war effort impacted positively on the price of whale oil, which increased from 14 pounds sterling a barrel on the eve of the war, to 26 in 1939, and 30 the following year, rising steadily to 45 pounds sterling in 1945. Compounding the effects of government demand, production remained in a lull, not least because most whaleboats had been converted into war vessels.

Beyond the world wars, the last reprieves in an overall trajectory of decline were granted by wars. The Korean War, for one, harbored fortuitous circumstances for the development of "pirate whaling." As China prohibited all oil-seed exports to the West when the war broke out in 1950, oil supplies began to thin out, thus ratcheting up whale oil prices. This provided the opportunity for a ship merchant such Aristotle Onassis, whose whaling enterprise was only one among many investments, to further inflate prices by sitting on his whale oil stocks. Average whale oil prices leapt from £99 per ton in 1950 to £144 the following year, while Onassis himself managed to obtain £172 (Tonnessen and Johnsen 1982, 523–525). Furthermore, Onassis had successfully placed his *Olympic Challenger* beyond any national jurisdiction by playing off one flag against another in the ship's registration. His catches thus went uncontrolled and unreported from 1950 to 1954. He had triggered one of the most destructive periods for whale stocks and left in his wake a practice that would continue to plague international efforts at regulating whaling. Two more bouts of pirate whaling occurred—first, in the 1960s (associated with the vessel *The Sierra*) and then in 1975–78 (linked to *The Tonna*, Day 1992).

National security concerns continue to prop up a dying Western whaling. The world energy crisis of 1973–74 saw whale oil prices heave momentarily. Furthermore, even as the demand for whale oil was petering out in the West, the interest in sperm whale oil was sparked anew by the development of sulfurization techniques, which could transform sperm oil into a high quality lubricant used in automatic transmission fluids and high technology applications.[6] One of the lesser products of the whale hunt had suddenly been rediscovered by the Cold War defense industry. From 1966 onwards, the world production of sperm oil exceeded that of whale oil. It became especially important for the USSR's space and military programs, as synthetic replacements were embargoed by the West (the United States and Europe). Sperm oil thus became a key component of the Soviet–Japanese whale trade relations. Into the early 1990s, the U.S. and U.K. governments were still maintaining stockpiles of sperm oil and whale oil to a lesser extent for national emergencies (Ellis 1992, viii), even while officially deploying policies against Japanese and Soviet whaling.

Whaling Nationalisms

The wars of the twentieth century illustrate the ways in which whaling was implicated in the strategic and defense considerations of the modern state. They reveal that whale products represented far more than an economic interest; they went to the heart of the national interest. Now the national interest, as Randolph Persaud (2001, 202) aptly emphasizes, "is made, manufactured on the basis of material conditions. In particular, [it] is understood as the official, authoritative and public expression of a way of life, that is to say, a culture." The material conditions of the modern whaling industry have been at the center of the analysis so far, thereby largely leaving aside considerations about the symbolic and cultural significance of these whaling societies. This section now turns to consideration of the ways in which whaling was involved in the formation of cultural complexes revolving around ideas of the nation and the national interest as they took shape in the interwar period.

Throughout the early twentieth century whaling had remained largely a bilateral affair, with Norway and the United Kingdom still holding between them 49 and 48 percent of Antarctic catches, respectively, until 1934 (Donovan 1992). With the opening of the pelagic era, these two countries saw their preponderance threatened by the rise of new whaling nations. The period is marked by the development of a vast interstate, nationalistic competition known as the "whaling Olympics," where all the whaling countries were pitted against one another to catch as many whales as possible before they could be taken by another country's whalers. Brought on by the opening up of the trade to new whaling nations in the mid-1930s, these dynamics continued apace for another two decades. This period also saw the first efforts to regulate whaling internationally (see chapter 4), and with them the threat loomed large of national quotas. These were not introduced until 1962, and only once it was too late—the whale stocks were too decimated to even fill them. The long, drawn-out international negotiations merely fueled the destructive drive for whales, because each nation's catches would be capped according to the country's share of overall catches. In the 1930s these "Olympics" drew out the whaling nationalisms especially acutely because they were bound up with the political buildup that led to the Second World War.

Japan and Germany were the two rising whaling nations in the 1930s, and in both countries, whaling was implicated in political economies revving up for war. In Japan, whaling had been mostly traditional, land-based, and localized. Japan's decision to enter into pelagic whaling in 1934 triggered a fundamental restructuring of its whaling operations and the emergence of the modern Japanese whaling industry. Pelagic whaling would play an integral part in Japan's military expansion in the 1930s and its integration into the circuits of international trade. Sales of whale oil to Europe (mostly to Germany) provided a valuable source of foreign currency to finance, notably, the war in Manchuria (Small 1971), and the vast quantities of whale meat that could be brought back from Antarctica once deep-freezing techniques were perfected (in 1938) suited Japan's food security objectives. Despite its late start, the Japanese whaling industry expanded with such speed as to cause considerable alarm among British and Norwegian competitors. Where British shipyards needed eighteen months to build a new floating factory in 1935, Japanese shipyards required four. Five new floating factories were converted or built between 1936 and 1938 (Tonnessen and Johnsen 1982, 418–419). Three years only after they began whaling in Antarctic waters, Japanese whalers were taking 11.6 percent of the whales caught (see figure 2.2). At the eve of the war, Japan's share was 19.2 percent, despite a pelagic fleet that still fell far short of either the Norwegian's or British's. From the Japanese perspective, the United Kingdom and Norway's newfound commitment to regulate whaling internationally, after decades (even centuries) of unregulated whaling, appeared as nothing more than an attempt to protect their own hands in the trade (see chapter 4). Hence Japan stayed away from these negotiations. The formidable expansion of Japanese whaling continued stealthily such that, by 1963, Japan had become the world's first whaling nation, with 41 percent of Antarctic catches.

Whaling had created new ties between Germany and Japan that were eventually sealed by a trade pact in 1939, whereby Japan committed its whale oil to Germany exclusively. One of the largest buyers of Japanese whale oil had been Unilever, whose profits were trapped in Germany in a complicated setup known as "the German fats plan" (Tonnessen and Johnsen 1982, 422–425). Hitler's policy of reconstitution

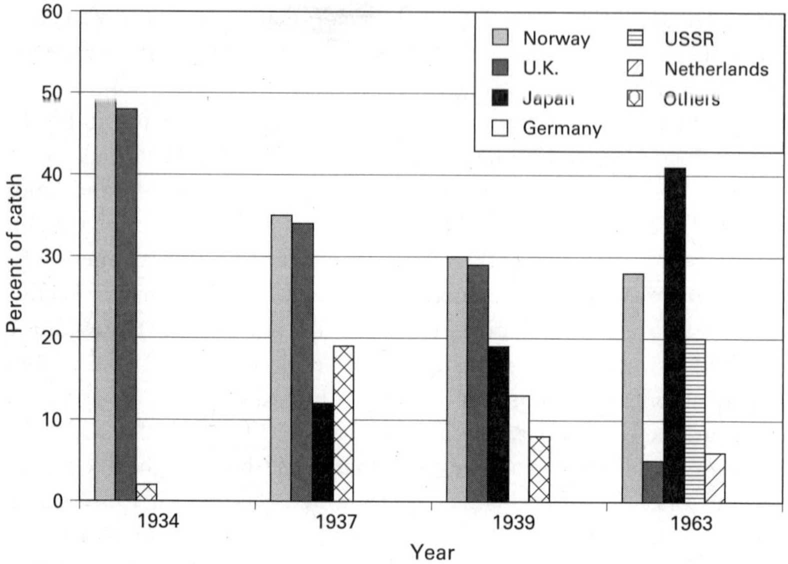

Figure 2.2
Composition of Antarctic catches (1930s–1960s). Data source: Tonnessen and Johnsen (1982).

of the German national interest meant protecting Germany against foreign interests while acquiring strategic raw material without spending any foreign exchange (which had been drained away the previous decade by the country's foreign debt). Whaling, it seemed, had a key part to play in Germany's policies of self-sufficiency. The success of margarine had posed a significant threat to German agriculture, a potent symbol of national interest in Nazi Germany. A 1933 decree restricted margarine's access to the German market, one of the largest markets in Europe. The problem was not margarine as such but that it should be produced on German soil by foreign interests (Unilever) with foreign (mostly Norwegian) whale oil. The importance to the German war plans of developing a modern German whaling feet was expressed by the military leader Hermann Göring himself, who remarked that foreign capital and Norwegian whalers "offer the possibility of supporting the supply of fats to our people, and thereby contributing to the attainment of the great goal of freedom in raw materials and food" (quoted in Tonnessen and Johnsen 1982, 398).

In the second act of the German fats plan, all earnings of Unilever's German subsidiaries were blocked and had to be reinvested in Germany. Unilever, who invested its profits in German shipyards as a result of this restriction, soon found itself effectively financing the building of a new German whaling fleet, in addition to buoying up Japanese–German whaling trade relations. Germany had thus embarked upon large-scale modern whaling by 1936.[7] By 1939, Germany was catching 13.3 percent of Antarctic whales (see figure 2.2).

The development of whaling nationalisms was not confined to the exceptional politics of the 1930s, nor were they dampened by the devastation of the war itself. In fact, they flourished with renewed vigor in its aftermath when the elimination of the third and fourth largest whaling nations was seen as a precious window of opportunity to enter into the trade by established and new whaling nations alike. We have already encountered the flurry of national whaling plans in the previous section. Let us examine in more detail here the Dutch whaling policies, as they provide a good case of postwar whaling nationalism steeped in whaling history.

Until the eighteenth century, the Dutch had dominated whaling. Since then, Dutch whaling had been on the decline, unable to withstand competition from successively British, then American, then Norwegian whalers. The plans to take up whaling anew had been hatched during the war by a group of businessman in a clandestine resistance movement (Tonnessen and Johnsen 1982, 523–535). The rationale was that, given that the destruction of the Dutch East Indies, effectively the country's granary, would require several years to repair, whaling would cover the country's immediate need for fats. The Dutch whaling plans were couched in terms of the country's autonomy and self-sufficiency, tainted in historical pride. At the end of the war the plan rallied government, monarchs, banks, and shipbuilding yards, in a vast national (re)building effort centered on whaling. A tanker was converted into a whaling vessel, and baptized after a figure of national glory, the explorer Willem Barendsz, who had given his name to the Barents Sea. The departure of the *Willem Barendsz* from Amsterdam in 1946 was celebrated as a national event. In 1950 the government turned down an offer by the USSR to buy out the floating factory well above its market worth, injecting instead vast subsidies to bail out this last remnant of a once glorious whaling fleet; it was eventually bought out by the Japanese in 1962 (and renamed *Nitto Maru*). In Great Britain, similarly, the importance of the fats supply to the country's national reconstruction was also cited as a justification by the British government for its interest in British whaling (Schweder 2000).

Conclusion: Back to the Power of Words

Whaling was an archetypal modern industry of the Second Industrial Revolution, comparable to other staple raw materials industries, such as coal. The modernization of its production methods through the use of new technologies marked the shift from a craft to a large-scale industry and distinguishes modern whaling from "traditional" and "opportunistic" whaling. Whaling was inscribed in the emergence of a "global system" (Sklair 1995) at the juncture where capitalist expansion, scientific innovation, and the global politics combined. Focusing on the consumptive and productive structures in which whaling was enmeshed served to show its pervasiveness, from the individual level, in the life of the early twentieth century modern consumer, to the level of the state. And yet a practice that had fed vital raw materials to the industrialization process vanished abruptly from the economic life of Western nations in the early 1960s.

In clearing the grounds for the subsequent analysis of the anti-whaling discourse, a central purpose of this chapter was to show that this discourse emerged in the West because *it could*; that is, that it was not prevented from doing so by a set of material interests aligned against it. What this illustrates is not the preponderance of material interests as an explanatory factor but rather the coconstitutivity of discursive and material practices. Indeed, the argument of this book is not that discursive power can overturn material power, nor that discourses are more powerful than material interests. Such an argument indeed would imply the possibility of disjoining discourses from material interests, such that they could successfully emerge *in spite of* material conditions aligned against it. Rather, the argument here is exactly the opposite: it is precisely because discursive and material interests are so deeply entwined that discourses cannot flourish without the appropriate material conditions.

Moreover, this chapter has shown that the configuration of material interests examined in this chapter does not explain everything. For the West's interest in whaling was still acute: even as the United Kingdom pulled out of whaling in 1963, three years later on the floor of the IWC it was still ferociously defending its right to resume whaling at a future date (Tonnessen and Johnsen 1982, 629). Even the United States, the whaling nation where whaling interests were most well and truly moribund, had systematically opposed the introduction of national quotas at the IWC from 1949 to 1962, because it too reserved for itself the possibility that it might once again engage in whaling (Tonnessen and Johnsen 1982). For whale produce had remained key to defense and national interest considerations, even after whaling had become uneconomical, as the involvement of the Dutch state has shown. Moreover, the prices of whale oil were steadily increasing, not collapsing. All of these factors sustained the West's continued interest in whaling, even after Western whaling had actually stopped on the ground. My contention is that whaling came to a halt in the West before it would have were it not for the anti-whaling discourse. Whaling could have resumed once the stocks had sufficiently recovered from overexploitation, as indeed it has done in Norway, and as these other Western governments also seemed to envisage at the time. Thus regarding the power of words, two main points can be drawn here. First, in this past whaling order, an exploitative whaling discourse sustained the Western interest in the practice beyond the life span of the practice itself in the West. In other words, this interest was maintained simply by the way the dominant discourse on whales and whaling ran. Second, it would take another dominant discourse, the anti-whaling discourse, to do away altogether with the possibility of that practice's ever resuming in most of those Western countries. What the anti-whaling discourse has done has been to make whaling inconceivable in the West today.

3
Whaling, Sovereignty, Governmentality

The previous chapter examined the role of whaling in the political economic order of modernizing nations during the twentieth century. It also served as a broad-brush overview of the wide range of whaling practices. This chapter examines more narrowly how whaling was integrated into particular structures of power and knowledge implicated in the formation of the modern nation-state. This requires digging deeper into whaling history, beyond the birth of modern whaling. In the process additional facets of the whaling nationalisms that were only roughly etched out in the previous chapter are drawn out. The chapter thus progressively unpacks the relationship between whaling and the state, inquiring into the specific areas where state and whaling practices conjoined. It explores how whaling was implicated in classic practices of sovereignty. Whaling served both to buttress sovereignty within, and to expand it outwards, through its involvement in the defense functions of the state and in its projects of territorial conquest. But it also played a part in the new cluster of managerial functions that characterizes the emergence of the modern state, for which Michel Foucault (2001) coined the concept of *governmentality*. The distinction between "sovereignty" and "governmentality" serves to provide two different angles for highlighting the connections between whaling and state practices. The final part of the chapter locates the emergence of cetology along this historical trajectory of the modern state. A brief genealogy of cetology serves to reveal the ways in which, from the beginning, the practices by which whales became "known" were constituted by a nexus of power/knowledge steeped in the national interest.

Whaling and the Emergence of the Modern State

The previous chapter highlighted how whaling became crucial to the nation in times of war. Whaling was not just an important component of wartime economies; it was directly implicated in defending the nation's sovereignty, as whaling fleets were systematically transformed into combat ships throughout both world wars. This was in fact old practice, which exemplified deep historical connections between whaling and the classic defense functions of the state. Throughout the nineteenth century wars in Europe, the British Crown regularly enlisted the whalers into its war efforts. According to the whaling historian Harry Morton (1982), whale ships were subcontracted as "Privateers." For these ships were especially sturdy; armed, they were particularly effective. They were private ships licensed by official permission of the state to capture enemy ships. Special prizes for these warring whaler captains were the whaling vessels of enemy nations and the whale oil they transported. Throughout the eighteenth and nineteenth centuries, all British shipmasters were required to swear an oath of allegiance to the King. Moreover, whaling helped to secure the state internally, as it was involved in the emerging police functions of the modern state. British whaling was directly employed in developing the colonial penal system: in 1786 the first convicts were transported to Australia on board two whaling ships subcontracted by the Crown (the *Lady Penrhyn* and the *Prince of Wales*). This became a regular commerce between the Crown and the whalers. The whaling enterprise Samuel Enderby and Sons' *Britannia*, for one, became renowned for its dual services, catching whales and transporting convicts. Thus within whaling itself surfaced two dimensions of state security—internal and external—which refer back to the two pillars of the modern nation-state, sovereignty and governmentality. Let us consider these more closely.

In observing the rise of modern society, Foucault observed how "the economic system that promotes the accumulation of capital and the system of power that ordains the accumulation of men became inseparable." A corollary was the emergence of "new technologies for the exercise of power" tailored toward "obtaining productive services from individuals in their concrete lives" (Foucault 1980, 132). This triggered the development of new forms of social organization that became increasingly invested by the state, such as discipline, which brought these lives into modern institutions (prisons, schools, factories) over the sixteenth to eighteenth centuries (Foucault 1975). That whaling was drawn into the deployment of the imperial disciplinary apparatus places it at the heart of processes that constituted the modern governmental state, a state that was increasingly involved in fueling and funneling the accumulation of riches. That is, the exercise of state power was no longer about securing a territory and the reign of the sovereign over its *people*. It was increasingly about managing a *population* as a productive resource. "Governmentality" thus designates the emergence of economic considerations into the political sphere—in other words, the rise of political economy. Foucault's broader concern was to capture facets of the modern state that had been overlooked by political theorists' narrow focus on modern guises of sovereignty. Hence the term was coined to provide a conceptual counterpoint to "sovereignty" and to highlight new ways of governing that were about directing the conduct of the governed from within—a manner of "ruling from within" rather than "ruling over." It serves to capture a state that behaves less as a traditional sovereign (displaying a show of force) and more like, say, a boat captain, or the head of the family steering the management of the household's resources (Foucault 2001). This shift is associated with the rise of new forms of productive powers that work through the very fabric of society, such as discipline and biopower (Foucault 1975, 2003). These modern forms of social power are not reducible to the state, even while increasingly invested by it. The concept thus broadens the focus beyond sovereignty and its top-down battery of laws to capture the bottom-up, social processes constitutive of the modern state. It draws out a series of continua rather than breaks between state and nonstate practices, public and private institutions, around an overall trajectory of increasing social power. Hence as Tim Luke reiterates in Foucault's wake, "The notion of governmentality invites social theorists ... not to reduce the complicated ensemble of modernizing developments to the actions of 'the state' [but instead] to investigate the 'governmentalization' of the economy and society" (Luke 1995, 27; see also Ferguson 1994). "Governmentality" is useful for teasing out the connections between a specific set of economic and social practices, such as whaling, and the various types of state-building practices brought into play around them.

Marshaled into wars of defense, whaling was entwined with the interests of sovereignty. However, it was also involved in the shaping of governmentality over precisely the period that Foucault considers in his historical analysis. For whaling implicated the state in the need to protect the national accumulation of wealth. British whaling, which had dominated the trade, saw its preeminence threatened by the rise of U.S. whalers in the eighteenth century, who were technically more advanced. American whaling had inaugurated pelagic whaling; by the mid 1840s, it would represent 70 percent of world whaling (Davis and Gallman 1993, 513). Thus at the outbreak of the War of Independence, Great Britain found itself importing almost four times as much oil from the American colonies as its own fleet could harvest off the shores of Greenland. During the hostilities, the Royal Navy drove the American whaling fleet from the seas, and from then on, the British government began deploying policies to protect British whaling. In 1775 Parliament voted to subsidize British-based Southern whaling through the levying of duties on American whaling products. It also offered generous bounties to entice the Yankee whalers away to the "mother country." In buttressing British whaling, the mercantilist state had thus developed its governmentality functions. At this point the Foucauldian lenses cast a particular light on the role of the state within the mercantilist system traditionally associated with the emergence of the nation-state in the sixteenth to eighteenth centuries, by highlighting the third term in the relation between the state and the economy, namely, the population, which tends to fade out of sight in most economic theories' analyses. It situates the state's interventionism within broader governmentality objectives of developing a particular relationship to the population wherein the latter is constituted as a productive force and harnessed into the nation's economic expansion. The subsidy of British Pacific whaling was justified in terms of national defense, illustrating the profound entwinement of sovereignty and governmentality. Whaling was seen as a naval school that maintained the contingent of skilled sailors for the Royal Navy, and thus a key resource for the naval power (Morton 1982, 96). These two streams—whaling and sovereignty, whaling and governmentality—run through the subsequent examination of state and whaling practices in the remainder of this and the following chapter.

Whaling and Territorial Expansion

Whaling was involved in the territorial deployment of sovereignty. For the capitalist dynamics captured in the previous chapter served the expansionist outlook of the modern nation-state. Whaling, it seemed, always had a part to play, from the games of strategic positioning on the sea routes of the globe in the early trade wars between the imperial powers to the last convulsions of a colonizing sovereignty that took place in the 1950s over the only unclaimed piece of land on the planet, Antarctica. This section retraces the role of whaling in shaping the outer limits of the modern state through some of these episodes.

Whaling was implicated in the constitution of naval powers. All naval powers had been, at one stage or and another, important whaling nations—Holland, Great Britain, France, Spain, Portugal, and the United States (Tonnessen and Johnsen 1982). For these imperial powers rivaling with one another, first over the sea routes of the globe and then over new territories, their whaleboats served as flagships marking their presence on the globe. This places whaling at the heart of imperialist expansion but also, given that the historical role played by these trade wars in the development of customary maritime law, at the heart of a nascent body of international laws. The drive toward the South Pacific illustrates the ways in which whaling was imbricated with the interest of colonizing sovereignties. During the second half of the eighteenth century, a series of naval wars between Spain and the United Kingdom had barred the South American coasts to British whalers. The whalers thus turned to the waters of the Southern Pacific, which they found teeming with whales, due to a combination of currents and migratory routes (breeding and feeding summer migrations). Yet they soon found themselves caught up in trade wars with the East India Company and in the long, drawn-out political effort to ease the country's monopolistic control over the waters of the East (Morton 1982, 95–96; Brading 1984), for these had remained the preserve of the Company, who had successfully maintained itself as sole licensing authority for British vessels on these seas, including whaling vessels. By the late 1780s the whaling lobby was pressing the British parliament to ease the Company's grip, and the 1789 Parliament Act opened the waters of Australia and New Zealand to British whalers, thereby unleashing the development of whaling in the region. Whaling thus drove the British colonization of the South Pacific. New Zealand, in particular, was formed as a settlement of whalers and sealers, to whom bringing law and order was a key justification for British sovereignty. British sovereignty, in turn, secured bountiful waters for the British whaling industry and thus a headstart over its main rivals, American and French whaling (Belich 1996, 127–140).

Whaling was also central to the colonization of Hawaii which, as Michael Shapiro (2002) and Sally Engle-Merry (2000) point out, occurred not as a swift political takeover but as a century-long enterprise driven first by merchants and missionaries, then by the demands of the whale fishery, and ultimately by the expansion of the sugar agro-industry. State institutions developed only subsequently and on the heels of private initiatives to lock in this economic appropriation by proclaiming it "American land." If one wrote the story of the making of the United States following the emblematic animal around which founding myth, territorial conquest, and economic expansion crystallized into a particular cultural formation, the narrative line would run from whales to cattle, via buffalos. Beef historian Jeremy Rifkin (1992) evokes the importance of animal-based produce in the emergence of a distinctly American economy. He shows the deep connivance between the military task of securing the land, its exploitation by the cattle industry, and the building of a transport system that linked the grazing lands of the Western range with the grain-producing regions of the east, which gave rise to the contemporary "global cattle complex." Moreover, the political project was explicitly linked to the evacuation of another culture's emblem, for the substitution of buffalo with cattle represented all at once the taming of the western wilderness and the subjugation of indigenous populations (Rifkin 1992, 82). The cattle were to the American prairie what the whales were to Hawaii. Furthermore, the shift from the whales to cattle signaled a major turn from an outward- to an inward-orientated economic expansion. The publication of Herman Melville's *Moby-Dick* in 1851 marked the cusp of American whaling and the beginning of its demise. From then on, the cowboy on land would come to replace the whaler on the sea as the figure of the pioneer at the heart of the U.S. identity.

Antarctica

Beyond the South Pacific, whaling sparked states' interests in Antarctica and remained throughout the first half of the twentieth century at the center of the battle for sovereignty over the last *terra nullius*. Whaling was instrumental in both discovering and charting Antarctica. The disputes as to who first sighted Antarctica in the year 1820 feature two British Navy captains and one whaling captain (the American Nathaniel Palmer). The regions of Antarctica were placed onto the emerging map by successive whaling expeditions. In 1830–32 a British Enderby Brothers expedition led by John Briscoe revealed the area around Enderby land, which includes the Adelaide, Anvers, and Biscoe Islands. The

Norwegian whaling Captain Horntvedt hoisted the Norwegian flag on Bouvet Island in 1927. Whaling was once again the motivation for the British proclamations of sovereignty over the entire Western Antarctica (South Georgia, South Shetlands, South Orkney, and South Sandwich). One incident in particular shows how an interstate dialogue taking shape around whaling was used to draw the contours of British sovereignty. The development of whaling fueled the interest in Antarctica and yielded increasingly acute territorial disputes, notably between the United Kingdom and Argentina. In the 1900s a Norwegian whaler, who sought to venture into the untouched waters around Head and McDonald Islands, asked the newly formed Norwegian Legation in London to investigate whether the Island belonged to the British and whether he should therefore apply to them for a license. Thus prompted, the British replied in the affirmative, thereby with their response effectively asserting their sovereignty over a piece of land whose status had hitherto remained indeterminate. Not only did the Colonial Office write a license for the Norwegian but it requested that the Norwegian whaling captain plant the British flag upon the island. The Norwegian foreign ministry took offence at the suggestion, and the matter was resolved by including a British subject in the expedition, who planted the flag himself (Tonnessen and Johnsen 1982, 182). The Norwegian whaling question had effectively compelled the British state into the position of sovereign.

This whaling history would be mobilized once again a few decades later as the territorial claims intensified in the "race to Antarctica" after the Second World War and prior to the treaty that suspended those claims in 1959. I draw here on a recently declassified 1956 CIA report, *National Intelligence Survey: Antarctica*, which was designed to formulate advice as to the United States' position on Antarctica. Produced by former whaling nations seeking to gauge how whaling played into their sovereignty claims over Antarctica, the text is useful for revealing how whaling featured in postwar interstate discourses.[1] What is interesting is that, although the report advocated that the United States relinquish all U.S. claims over the region in favor of an international treaty that would better serve U.S. interests, it was still intent on disputing with the British claims about who discovered the land. The CIA identified whaling nationalism as a key driver of British Antarctic policies. For this former whaling imperial power, who saw its sovereignty increasingly shrinking with decolonization, Antarctica featured in as the last piece of land upon which sovereignty could be asserted. The CIA concluded:

Concern for national prestige and sensitivity arising out of many enforced post-war retrenchments and losses of power partly explain why the protection of British interests in Antarctica is important and why the government and press have sought to enlist support for a vigorous Antarctic policy (CIA 1956, 20).

In such a context, whaling expeditions to Antarctica came to be seen as "British exploits" and the whalers themselves incarnated "British heroism and love of adventure," such that "No U.K. government could afford to ignore this public pride and interest" (CIA 1956, 20). The economic importance of whaling is then assessed in the following terms:

Whaling (and sealing to a much lesser extent), though of diminished importance, remains the major economic incentive. Schemes for exploitation of the area for its mineral "wealth," or as a base for the development of commercial air routes, or for harnessing Antarctic winds for electric power have had almost no appeal. Within the fisheries, whaling is by far the most important element (CIA 1956, 20).

The document then lists most of the uses of whale produce we encountered in the previous chapter.

Another interesting point that surfaces in the report is the ways in which former whaling colonies of the South Pacific mobilized this colonial whaling past in seeking to extend their still relatively new sovereignty over Antarctica. The document remarks that the "importance of whaling to the world was cited as justification of Antarctic claims of Australia and New Zealand" (CIA 1956, 20). Regarding New Zealand, it stipulates, "Although the economic importance of the Antarctic is regarded as highly speculative, in New Zealand the importance of whaling is recognized" (CIA 1956, 36), such that the government was coming under increasing criticism in the late 1950s for failing to develop its whaling policies:

In recent years the New Zealand press has criticized the government for neglecting one of the greatest wealth-producing whaling areas in the Antarctic. Support for a vigorous program to exploit these resources has come from a number of Members of Parliament who are concerned about the present depletion of the fish in New Zealand coastal waters and the growing demand for fish and whale meat in other countries (CIA 1956, 36).

Yet less than a decade later the New Zealand parliament would launch among the most vigorous anti-whaling policies, as we shall see in chapter 7. As for Australia, soon also a staunch anti-whaling state, in the 1950s whaling was fervently resumed in the interest of sovereignty:

[Australia's] Interest in Antarctic whaling has been particularly marked when the southern seas have been "invaded" by the whaling ships of other nations, especially those of Japan and the USSR. With the sudden increase of interest in Antarctica in 1946–47, Australian scientists and politicians linked whaling with the need of launching scientific expeditions to bolster claims. Official plans were laid for a coastal whaling industry for Australia proper, which could lead to Antarctic whaling when ships were available (CIA 1956, 53).

The document then provides a detailed description of the expansion of Australian off-shore whaling from five whaling stations (three on the West coast, two in the East), for which "a Norwegian adviser was obtained for the Australian whaling industry." Whaling was thus firmly lodged in the nexus of "economic, strategic and scientific" interests shaping Antarctic policies. Also tangible in this last abstract are the connections between the science of whales and whaling politics that I now turn to consider more closely.

Cetology and the National Interest

Another contribution of Foucault's work has been to draw attention to the deep connections between the exercise of power and the accumulation of knowledge (Foucault 1975). This coconstitution of power and knowledge transpires through the establishment of whales as an object of scientific research. In retracing the birth of cetology, this section focuses on two aspects: the type of knowledge-producing practices through which cetology was constituted as a scientific discipline and the structures of power within which these practices developed. The knowledge practices, first, are key to understanding the way in which the whales were perceived. They formed the knowledge component of whaling discourses. They prescribed the ways of talking about the whales, insofar as discursive practices always presuppose a field of knowledge. In Foucault's words:

> Discursive practices are characterised by the delimitation of a field of objects, the definition of a legitimate perspective for the agent of knowledge, and the fixing of norms for the elaboration of concepts and theories (Foucault 1977, 199).

Second, to tease out this nexus of knowledge and power—in both its economic and political guises—I examine successively the relationship between science and the whaling industry, and science and the state. This triangular relationship was essential to operating the shift from a sporadic and disparate set of empirical observations about the whales to an institutionalized science. It was, to use Foucault's term, central to the "knowledge of whales" becoming *epistemologized*.

The science of whales was spurred by the expansion of whaling. For insofar as understanding whales was necessary to catching more of them, it was an important aspect of the whaler's craft. The first insights into whale behavior were drawn from the notes of whale captains such as William Scoresby, an important figure in late eighteenth century British whaling, who jotted his observations alongside the catch records in his navigation journals (Matthews 1975, 169–182). The takeoff of cetology as a scientific discipline was coextensive with the rise of the modern whaling industry. Whale research was part and parcel of modern whaling, for the industry both funded the scientists and hosted their activities on board. The research itself was conducted directly on the decks of the factory ships, where the scientists would measure, weigh, and examine the animal using the same instruments with which it would be processed into oil. The scientists also held the logbooks of the whaling companies. Industrial processing and scientific analysis were thus intertwined in this era of early "knife and notebook" cetology (interview with Ray Gambell). This also determined what type of research was carried out by a "slice and dice" science, as it was also sometimes known (interview with Per Palsboll): pregnancy and embryo studies, food studies, and age, all of which required the whale to be cut open in order to inspect its internal organs. In contrast with today, where lethal research methods are a major subject of debate in the scientific body of the IWC (Epstein 2005), such practices were considered acceptable forms of knowledge accumulation. Indeed, keeping the animal alive was simply not a concern for the scientists. The cetologists studied the whales through the same lenses as the whalers. Furthermore, they remained dependent on the latter both for data and employment. Far from developing as an independent field of research, the science was thus heavily bound to the whaling industry, both in its funding and methods and even with respect to the contents of research. The science of whales was harnessed to the expansion of the whaling industry.

Scientific interest was also what brought the United States back to Antarctica, long after the economic interest in whaling had petered out. "In the 1920's," noted the CIA (1956, 62), "there began [in the United States] a series of well-organized expeditions for scientific exploration sponsored both by the government and by private organizations and individuals." This coincided with the setting up of a series of U.S. national research programs where the state's interest in whale research was commensurate with its interest in Antarctica, although in this case it could seem still relatively untainted by political interest since the United States had held back from staking any claims over the area. In the United Kingdom, the conflation of scientific and political interests was more immediate. As early as 1842–43, the British Admiralty sponsored a scientific expedition to Antarctica led by Captain Ross. In 1924 the Discovery Committee was set up under the secretary of state for the colonies "to survey the marine resources of the Falkland Island Dependencies" (Matthews 1975, 172). Yet the research that ensued was not a comprehensive research program that considered all marine resources equally for their potential benefits to science. Rather, launched by the Colonial Office in 1926, and funded by the taxes levied on the whalers, it focused on whales alone, which were by far the most economically beneficial resource. Furthermore, the chairman of the Discovery Committee, C. R. Darnley, was also the whale licensing authority over Antarctica. The Committee maintained ongoing research programs that were only briefly interrupted by the war. Annual expeditions were carried out with two ships, *Discovery II* and *William Scoresby*, with the British cetologist Neil Mackintosh at their head. Government-funded science continued to operate in tandem with the whaling industry, with, for example, the various whale-marking schemes, where whalers were rewarded for returning the serialized steel dart used to mark the whale (Matthews 1975, 180).

Furthermore, these scientific expeditions effectively relayed the whaling industry in charting the waters of Antarctica, illustrating the profound connivance between the modern tasks of *knowing* and *owning*. "The work of the committee," reported the CIA, "whose object was to investigate the life history of Antarctic whales, resulted in considerable exploration of the peninsula as well as the mapping of large parts of the South Sandwich, South Orkney, and South Shetland Islands, particularly by the expeditions of Drs. N. A. Mackintosh and Stanley Kemp" (CIA 1956, 31). Once again, science and the national interest were fundamentally entwined, as all of these subsequently became British territorial claims. The Committee's exploratory work reached its peak in the years 1934–37, which were also years of intense pelagic whaling as well as increasing government involvement in whaling regulation, as we shall see in the following chapter. These expeditions were sponsored mainly by the Colonial Office, the Admiralty, and War Office (CIA 1956, 31). The scientific and military functions seemed ever more densely interwoven after the war when the Committee acquired highly efficient Navy corvettes for its scientific research. Thereafter military technologies developed during the war, such as radar and sonar, were increasingly applied to the study of whales.

After the war the Falkland Islands Dependencies Survey (FIDS) was created to pursue the exploration of Antarctica. To the CIA it was merely a scientific front concealing "political and military" functions: "While the principal purpose of FIDS is reputedly its exploration and research, there is considerable evidence that it was designed primarily as a means of maintaining the United Kingdom's claim to sovereignty" (CIA 1956, 31).

In fact, this connection between the interests of science and of sovereignty had always existed and had served to constitute them both. Moreover, once the scientific and military functions began to be separated, whale research slowly petered out in the United Kingdom. After the war whale research proper was transferred to the National Institute of Oceanography (NIO), where it was integrated with the study of oceans as a whole. Thereafter little new British research was conducted, although the NIO did retain an important coordinating role in the emergence of international whale research programs (CIA 1956, 31). In 1976 cetology was moved yet again to the Natural History Museum, under the auspices of the Marine Mammal Working Group. Yet as British cetologist and first IWC Secretary Ray Gambell recalls, the interest in whales was lost, and the whale unit was simply a spin-off of the work done on seals. By the late 1970s, there was only one person working on whales in the museum (interview with Ray Gambell). While scientific research had by then largely been taken over by the IWC, it continued to depend on research conducted by each member state.[2] In the United Kingdom, the government's interest in whale research programs expired together with the whaling industry.

This chapter has shown how whaling was involved in the making of the modern state, both in drawing the contours of its sovereignty without and in its emerging forms of governmentality within. It then considered the emergence of the science of whales as the "regime of truth" pertaining to the dominant whaling discourse. It thus teased out the matrix of state, science, and industry relations underpinning the whaling order, and the particular structures of power/knowledge within which whaling was normalized at the national level. The following chapter considers how these structures of normalization extended to the global level.

4

The Society of Whaling States

After featuring whales as an object of everyday life, and as an object of national interests, this chapter considers whales as the object of an interstate discourse that effectively posited states as the subjects of international cooperation. For the first dominant, global discourse on whales was a state discourse that took shape around the necessity to regulate whaling. Discursively, the shift from the national to the international level operated around whaling as an object of regulation, which crystallized the emergence of international cooperation. This passage is at the heart of this chapter. It examines the emergence of an international whaling order by following the trajectory of whaling regulations from the first efforts by Norway and the United Kingdom in the late 1920s to the IWC that was eventually created in 1946 and is still today the institution in charge of whaling governance. However, as international regulations evolved out of provisions already in place in certain countries, national regulations provide the starting point for retracing the emergence of state regulatory and discursive practices around whaling. At this level the analysis continues to unveil the connections between whaling and the state. Attention is paid to the ways in which the development of whaling governance served to constitute the identity of certain states as "whaling nations," both internally and vis-à-vis other states. The chapter then analyzes the formation of a "society of whaling states" at the international level, deploying the conceptual tools of social theorist Pierre Bourdieu.

The broader theoretical purpose of the chapter is thus to examine the social dynamics of interstate cooperation. States engage in collective action not always as a result of a clear-cut, utilitarian, cost–benefit calculation but sometimes because they have been normatively inclined toward it by the ways in which they have been socialized into the society of states. This is not to say that states are coerced into cooperation nor that they would act against their interests. The state system is, after all, an "anarchic" system. Rather, it is simply to emphasize that international cooperation occurs within a dominant discourse where a particular normative order has become entrenched and subsequently ordains the ways in which states behave with one another. Thus, in some instances, focusing on the dominant discourse within which the collective action has been devised is more pertinent than a painstaking calculation of the costs and benefits that led each state to partake in that collective action. In our whaling case, a strict cost–benefit assessment of why states voted to suspend commercial whaling at the IWC in 1982 would not say very much. While the costs may have been insignificant for a majority of countries that no longer whaled, the analysis would be hard-pressed to express these states' whale-saving interests in terms of quantifiable, tangible benefits, especially since they had been brought together in 1949 by exactly opposite interests (exploiting whales rather than saving them). Rather, a majority of states voted for the moratorium because they subscribed to a new, moral notion that destroying whale stocks was wrong, which, in turn, presupposed a common system of values and beliefs. In other words, it implied a common discourse about whales—hence at least some basic form of social structure. The remainder of the book is devoted to analyzing how this discourse arose and subsumed the previously dominant whaling discourse. The main point here is that social dynamics have remained a key explanatory factor of state behavior with regard to the whaling issue, notwithstanding the shift from a society of whaling to a society of *anti*-whaling states (see chapters 7–10; see also Epstein 2006).

Moving Away from Regime Theory

As this focus on social dynamics requires a significant shift from where it has conventionally sat in the examination of international whaling politics, namely, away from regime theory, it is important to first locate the point of departure with these approaches (Friedheim 2001; Andresen 2001; DeSombre 2000; Mitchell 1998; Stoett 1997; Peterson 1992). For this, regime theory's basic account of international cooperation is useful. Because cooperation is costly, states will not interact with one another without a good reason, especially when they share no common borders, which was the case with most whaling states. In fact, the whaling case appears to confirm the so-called neorealist emphasis on noninteraction (or rather noncooperation) as the default condition of the international system. The odds were heavily stacked against the chances of whaling's ever being contained by collectively devised international regulations. The industry was hardly inclined to self-restraint as we saw in chapter 2. In fact, whaling has often been held up as the paragon of the commons' tragedy, where the inability of economic actors to move beyond short-term interest maximization leads to the irreversible depletion of the resource upon which they depend and, eventually, to their own as well as the resource's destruction (Mitchell 1998; M'Gonigle 1980). As for states, as they were becoming increasingly enmeshed in the competitive dynamics of the "whaling Olympics," they too had scant incentive to limit their own whalers for the same reason that every whale uncaught added to another nation's catches. Indeed, the explanation continues (in characteristic eco-authoritarian fashion à la Terry Hardin–William Ophuls), because whaling occurred beyond the remit of territorial authority, it eluded the state, the institution indispensable to containing this short-sighted, selfish behavior. Hence states were not especially disposed toward joining forces with other states to limit their whaling, as indeed the difficulties encountered by the first efforts to regulate whaling will show.

Yet international whaling regulations did emerge, albeit belatedly. If they were so costly, why did they develop at all? Was it because, as whales were becoming harder and harder to catch in the face of depleted whale stocks, a tipping point was reached where states finally realized that the costs of not regulating outweighed the costs of regulating? International regulations would indeed share out the cost of limiting catches between states, and keep both the whales and the industry alive. In a typical neoliberal institutionalist account, they made sense as a cost–benefit calculation (although they ended up being adopted too late to be optimal). However, I suggest, there were other, social, costs that began to be incurred within the society of whaling states that was taking shape, new types of peer pressures pushing states toward developing collective regulations. Furthermore, as these social dynamics consolidated, there began to emerge new types of benefits to cooperating. These types of costs and benefits cannot be comprehended from a perspective that envisages states exclusively as unitary, atomistic actors. This is the limit to both neorealist and neoliberal institutionalist variations to regime theory (see notably Haggard and Simmons 1987; Young 1989a, 1989b, 1990, 1999a, 1999b; Antunes 2000) and to narrow cost–benefit calculations that ignore their social dimensions.

The approach proposed here does not deny states their agency, nor even that they are self-interested. However, in devising their interest-maximizing strategies, states also take into account these costs and benefits that have to do with their belonging to a social system. States do not come to the cooperating table with preconstituted interests or wills that are fixed once and for all.[1] Rather, their wills and perceptions of their own interests are always in the making, and they are also (but not only) shaped by their relations with other states. States want to belong, not because of an inherent attribute of their "statehood." Rather, while there may be instrumental reasons and material payoffs to positioning oneself in a particular way on an issue (as "all for saving whales," in the case of the 1982 moratorium), to presume so requires a discourse in which it is already possible to behave as a whale-saving state (see chapter 7). Therefore, only when states have stepped into this subject-position do they adopt the interests that they do. Thus membership in international organizations such as the IWC—belonging in a social sense—is necessary to make certain interests possible. States need to belong to have state interests—hence the need to envisage a different kind of, social, strategizing at play in the constitution of state interests and identities.[2]

Toward a Bourdieusian Analysis of Interstate Interactions

Social dynamics play out within a particular discursive order, or *nomos* to use Pierre Bourdieu's term, which transpires in all the discourses and practices around whales.[3] How, then, were states convinced to join the regulatory efforts? By being socialized into a particular space of social interactions, I argue. "Socialized," here, is understood as the process by which actors are drawn into a particular position within a social field from which they interact with other actors.[4] Hence the need to look at how this field of interactions was initially laid out, as it accounts for the particular ways in which states *can* (but not necessary will) behave with one another. The concept of "field" used here is that of Pierre Bourdieu, which, together with his analysis of interests, is especially useful for thinking through the relationship between the structure of the international system and the agency of states interacting with each other beyond the simple juxtaposition implicitly conveyed by the classic (neorealist) metaphor of the international system as a billiard board. First, let us consider Bourdieu's concept of "field" (see also Guzzini 2000, Holzscheiter 2005). *Fields* are structured spaces of differential positions from which social actors interact with one another and develop relations (Bourdieu

2002a). Although Bourdieu originally devised the concept at the individual level, it extends to other types of social actors altogether—such as states—since Bourdieu himself (2002a) worked to draw out the common properties of social agency within a broader general theory of social fields. The key feature of the concept is that it foregrounds the competitive dimension of social interactions yet *without* reverting back to the atomistic, "grab as much as you can and run" neorealist model of competition. For competition is an (perhaps *the*) inherently social process—it presupposes socially significant others to compete with and, indeed, enduring relations with these others. Bourdieu, whose thinking often runs along sporting metaphors, approached the space of social interactions as a *playing field* on which the social actors are constantly competing with one another for gains that are specific to that field. Although Bourdieu has tended to focus on mature social fields, one text in particular (Bourdieu 2002b), in which he analyzes the emergence of sporting practices as a social field, is especially useful for looking at the emergence of whaling regulations. There he identifies the point at which a system in which the actors and institutions were merely juxtaposed (much like in the billiard board model) turns into a *social* system as the moment when the competitive dynamics kick in. Another important feature of "fields" is that socialization is not only competitive but it is an unequal process. The conflict over the gains specific to that field between the actors standing in dominant and dominated positions is an organizing principle of competition in that field. This is an important departure with the Habermassian-inspired constructivist perspective on socialization, which has tended to treat it as a neutral, cooperative process, grounded in consensus (Checkel 2005, Risse 2006). These differentiated positions thus endow the respective actors who occupy them with different means of strategizing. It endows some more than others. However—and this is crucial, as it restores the role of agency—this does not mean that the actors' game is entirely predetermined; it simply indicates the means with which they start off in the game.

Second, the ways Bourdieu articulates "agency" and "interests" breaks open the individualist–utilitarian monopoly on the concept of interests by replacing them within a sociological perspective. In so doing, he offers a much richer understanding of what it means for an actor to belong to a social system—or rather how this belonging reorients its behavior. An actor's interests are not predetermined; they are constituted within a social field. Bourdieu defines interests as a "specific investment in the stakes" of a particular field (Bourdieu 2002b, 119). This investment, he continues, is a both the necessary condition for belonging to the field and the product of this same belonging. This captures the agency of the actor: it is up to the actor to make that investment, and it can potentially choose to hold back. Indeed, whaling states can (and indeed often have) withheld from partaking in the collective negotiations at the IWC. However, once it is made, the actor then belongs, and the mechanisms of this belonging become central to the actor's perception of its own interests. That is, the actor is tied to the social system by its very interests: it *wants* to belong. Thus it will tend to play by the specific rules, to talk the particular talk, to observe the *nomos*. In addition, the actor is roped in by the common interests that kick in. This is another important advantage of Bourdieu's redefinition of interests: it draws out a commonality of interests. Bourdieu shows how social actors, even while they are competing with one another to maximize these (socially constructed) interests, share some interests in common: those that have to do with the field itself and with belonging to it. All actors who position themselves in a particular field want the field to continue to exist, because as it defines them, it has become the site of their interest-maximizing strategies. In other words, there is an interest in maintaining the field, because the field defines their interests, and therefore their identities.

In light of this discussion of fields and belonging, this chapter will show how early attempts at regulating whaling constituted a social space that served to normalize state whaling practices. It begins by retracing the trajectory of whaling regulations from early attempts at bilateral cooperation to the first halting steps at multilateral regulations that eventually yielded a full-fledged society of whaling states with the creation of the IWC. Belonging to the IWC became, for states, a way of marking themselves as whaling states and obtaining recognition from other states for this whaling identity. What is interesting is that these very same social dynamics would then be mobilized, a few decades later, to transform the IWC into a society of *anti*-whaling states. They thus warrant close examination.

Bilateral Regulations (1906–1930)

Whaling was inherently difficult to regulate. Involving expeditions onto the high seas, it operated beyond the remit of territorially defined state controls. The space of whaling had always been aterritory, a space of alegality. This quality was in fact critical to Melville's use of the whaleboat as a space of lawlessness and thus, paradoxically, as the perfect platform from which to critically appraise the laws of the land. As the new technologies of modern whaling drove it ever further from the coasts, and eventually to Antarctica, it only further escaped scrutiny. Yet as the industry grew, states progressively realized the need to control it. Thus running through the rise of modern whaling was an inherent tension between an expanding industry and the state's efforts to contain it. Of course, the relationship between the state and the whaling industry varied greatly from one country to another. At one end of the spectrum, the American whaling that inspired Melville was a case of completely unregulated whaling, and it is studied by sociologists precisely on this account—as a case of a lawless, stateless society (Creighton 1995).[5] However, it also remained highly localized, concentrated at 80 percent around the shores of New Bedford, Massachusetts, until it progressively petered out (Davis and Gallman 1993, 513). U.S. whaling never took the turn toward large-scale modern whaling described in chapter 2. Thus while whaling was highly significant in shaping local, coastal identities—it was central notably to the Yankee identity—it never became a defining feature of the *national* identity. This, I contend, is because of the lack of the federal state's involvement. As we saw in the last two chapters, whaling nationalisms implicated the state as a prism–organizer of the national identity. This is what accounts for the United States' having been all at once one of the most prominent whaling nations materially and one where whaling never took hold symbolically as a marker of the national identity.[6] At the other end of the spectrum lies Norway, whose identity as a whaling nation was sealed not just by the spectacular rise of its whaling industry but by the development of a regulatory framework to contain it. In controlling whaling, the state was also deploying itself; it was both developing its own regulatory powers and asserting a Norwegian whaling identity. There whaling regulations were coextensive with the building of the modern state. In Norway, the management of whales and whaling is what effected the rationalization of resource use and the creation of bureaucracies to manage nature (Murphy 1994). Thus in Norway a similar process to what occurred in the United States with regard to forests (Knobloch 1996) took place around whales.

Any attempt at regulating whaling had to tackle the developments in Antarctica. In 1914, Antarctic production yielded more than half (54 percent) of the world's whale oil, and almost all of it (92 percent) by

1938 (Tonnessen and Johnsen 1982, 313). As a result, given Antarctica's unclaimed status, whaling regulations would necessarily involve some form of international collaboration. On the other hand, Antarctic whaling was still relatively concentrated in the hands of British and Norwegian whalers, who together held 97.5 percent of the Antarctic production (see figure 2.2). Thus bilateral cooperation between these two countries would effectively contain most of the world's whaling. Both countries devised strategies to control whaling grounded in their respective strengths in the region: the United Kingdom using its territorial presence in the region, and Norway its monopoly over skills and materiél. The international regulations of whaling emerged at the cusp of these two modes of regulation.

The British Licensing System

With the issuance of the Whale Fisheries Ordinance in 1906 by the Governor of the Falkland Islands, the United Kingdom imposed a system of licenses and permits to unilaterally control whaling in the Antarctic waters. Enforced from the land base in South Georgia, it included a complex range of royalties, export duties, and penalties which brought, in the CIA's analysis, a "valuable revenue" to the British Antarctic administration (CIA 1956, 20). The first license, granted to a Norwegian for the year 1906–1907, established that all whales, irrespective of whether they were caught close to South Georgia, were to be regarded as caught in British colonial waters (Tonnessen and Johnsen 1982, 181). Claiming ownership over the whales, the British state thus claimed ownership over the waters and progressively extended its sovereignty over the territories of Antarctica. For as more and more applications came in from Norwegian whalers in increasingly remote areas, the British authorities continued to issue licenses for areas whose sovereignty was still undetermined (Tonnessen and Johnsen 1982, 339). The license application became an integral part of Antarctic whaling practices; thus it contributed to legitimizing whaling, inasmuch as regulation also constitutes a mode of normalization.

The licensing system would be subsequently mobilized in the territorial disputes with Argentina, in which the British argued that the licensing process expressed a tacit Argentinean recognition of British sovereignty (Tonnessen and Johnsen 1982, 178). Moreover, it was directly involved in the competition that developed between British and Norwegian whaling and in the rise of the British whaling nationalism (see chapter 3). The

British interest in whaling, and its whaling identity, developed in reaction to the success of Norwegian whaling in Antarctica.[7] For the first five years, applications had originated mostly from Norwegian whalers, who had opened up the Antarctic waters to whaling. Their success triggered an "urge to raise the British flag" among British whalers (Tonnessen and Johnsen 1982, 338), who demanded (unsuccessfully) that the licenses granted to "foreigners" be transferred to British whalers and reserved for British whaling. Thus from a narrow territorial basis, the British had developed a regulatory system which instilled some measure of regional, if not yet international, control over the most important whaling grounds that would have remained otherwise completely uncontrolled. Furthermore, they carved out for themselves a position as whaling regulators which was at odds with their real place in the whaling industry, a trait which was to be maintained throughout the emergence of whaling regulations.

This territorially based system of controls remained relatively enforceable so long as the floating factories had to return to the whaling stations concentrated mostly in South Georgia to refuel and to process their catches. However, the rise of pelagic whaling after the introduction of the stern slipway in 1925 accrued the Norwegian whalers' operational autonomy, since the whales could be fully processed and stored on board. Recognizing this, the Colonial authorities pressed the Norwegian state to control its whalers directly. The development of Norwegian whaling regulations effectively operated the shift from a territorial to a nationality-based regulatory system.

Norway and the Regulation of Whaling

The rise of a modern, regulated whaling went to the heart of Norway's establishment as a modern, technologically advanced, independent state. Norway had been the birthplace of the modern whaling industry and the country that had taken factory whaling to Antarctica. Whaling was significant as one of the few domains over which this young nation held an uncontested lead over the other nations throughout the first half of the twentieth century. Indeed, even after Norway had lost its position as the first whaling nation, Norwegian gunners and technology remained highly sought after. This monopoly over skills provided a de facto means of control over the industry worldwide—so much so that when Norway finally enacted its Crew Laws in 1945 (after threatening to do so for over a decade), by which it outlawed Norwegian involvement in foreign whaling, it succeeded in reigning in the postwar resumption of whaling (Tonnessen and Johnsen 1982, 521–527). Upon its accession to independence from Sweden in 1905, modern whaling became a symbol of national pride around which the nation could build itself, both internally and in relation to other states. In the analysis of Norwegian whaling historians Tonnessen and Johnsen (1982, 71):

> The creation of modern whaling was a Norwegian national triumph, enhancing the Norwegian people's sense of self-reliance and confidence, turning Svend Foyn into a national hero and his enterprise into a victory for Norway.

Whaling was rapidly implicated in the shaping of a Norwegian foreign policy (Arlov 1993). Indeed, the early days of Norwegian independence had been marked by an increasing discrepancy between a rising Norwegian presence on the world's whaling grounds and the political weight of the country in the international system. Norway came to the 1920 Paris conference, the very conference that founded the League of Nations and thus the postwar international system, as the first whaling nation, and whaling was put forward as grounds for obtaining sovereignty over the island of Svalbard, where Svenn Foyn had tested the first factory ship. Yet this was also a regulated whaling. In fact, Norway's attempts at regulating whaling effectively predate its establishment as a sovereign state, since the first measures date back to 1904, with a ban placed on whaling off the coast of Finnmark to protect fishing interests, the same ban that had provided the initial impetus for the exploration of Svalbard. Norway's interest in regulating whaling was commensurate with its interest in the industry: precisely because whaling was vital to the national interest in so many ways, it was vital to regulate it to avert the depletion of the resources and thus the extinction of the trade.

At the same time, the development of Norwegian whaling regulations was also an outward-oriented process, in that they emerged in response to growing international concerns. The need for whaling regulations was voiced by the League of Nations as early as 1925, the very year pelagic whaling took off in Antarctica. At the first international conference on the topic convened subsequently in Paris in 1927, the Conseil International Pour l'Exploitation de la Mer thus declared:

> The riches of the sea, and especially the immensity of the Antarctic regions, are the patrimony of the whole human race. To save this wealth, which, being today the uncontrolled property of all, belongs to nobody, the only thing to be done is ... to base a new jurisprudence ... on the scientific and economic considerations which, after all the necessary data has been collected, may be put forward, com-

pared and discussed at a technical conference by the countries concerned (quoted in Donovan 1992).

This signaled the irruption in the international regulatory discourse of a constitutive tension between the need to conserve whale stocks and to develop the industry that would run through all regulatory texts and resurface in the 1946 Convention (see chapter 6). By 1929, the first movement of scientists to initiate the protection of dwindling whale stocks was afoot. Six American zoological associations set up a Council for the Preservation of Whales. No international regulation resulted from the Paris meeting. Only Norway, pressed by the British, took action and passed what was characterized by the scientists as "the most constructive legislation ever drawn up to save the waning animal life" (Tonnessen and Johnsen 1982, 401). The 1929 Whaling Act endowed the Norwegian government with powers to regulate its whalers greater than any other nation. It subjected the whalers to a system of taxes and tight controls. It comprised set whaling seasons, an inspector scheme that placed a government-appointed inspector on every factory ship, who would record every catch in a mandatory logbook. For the first time, a minimum length for each species was set, and the catching of the already endangered right whales, as well as calves or females with a calf, was outlawed.[8] Nor did the state hesitate to use these powers: in 1931, for example, the whaling season was called off entirely (Tonnessen and Johnsen 1982, 366).

The 1929 Whaling Act was perhaps more significant in terms of creating new institutions and shifting the norm of whaling practices than it was in terms of actually protecting the whale stocks from destruction. In the Norwegian context, it triggered the development of the state bureaucracies of natural resource management, and it inverted the relations between the state and the industry. The law established the first Whaling Council (Hvalrad) as an advisory board to the government. It was initially composed exclusively of scientists and members of the Ministry of Industry and Foreign Affairs and only later allowed one industry representative, who was, furthermore, barred from the chair. Furthermore, science was established as a cornerstone of this regulatory system, a feature that significantly impacted the design of an international whaling regime. The State Institute for Whale Research (Statens Institutt for Hvalforskning) was created to provide the Whaling Council with scientific advice, based on research programs conducted in cooperation with whaling companies. Most importantly, the law established the separate

Bureau for International Whaling Statistics (Komiteen for Internasjonal Hvalfangststatistiskk) in Sandefjord, to which the whalers were to report back their catches after each expedition. The Whaling Bureau compiled all data into the annual *International Whaling Statistics*. This had two key effects, on scientific practice and on science's institutional autonomy. Creating a separate scientific institution was key to establishing cetology as an authoritative, independent body with the monopoly over knowledge and truths, an important institutional feature of the IWC as we shall see in chapter 7. As for the scientific practice, it inaugurated the statistical approach to the study of whales, pioneered by the scientists Hjort, Lie, and Rudd (the latter a future IWC cetologist), an important contribution to whale behavior analyses (Matthews 1975, 172). The *International Whaling Statistics* still constitutes the main source for studying the evolution of catch histories, which remains a cornerstone of whale management today.

The First Whaling Conventions: A "Presocial" Field (1931–1937)

The First International Whaling Convention

The scientific developments prompted by the Norwegian International Whaling Bureau served to underscore the urgency of controls. The science here is used differently than in the previous chapter, where it was approached "from without" and in relation to the political context where it developed. Here, I look *through* the scientific lenses to see what they revealed to the regulators at the time. While the exact damage to whale stocks is (still) notoriously difficult for scientists to gauge exactly, two figures could provide a measure of the industry's excesses, catch numbers and catch composition.[9] Catch records doubled with the introduction of the stern slipway, leaping from 7, 271 whales (all species included) in 1923–24 to 14, 219 whales in 1925–26 in the Antarctic alone (Aron 2001). They then steadied slightly but surged to 20,341 whales in 1929 and then again to 30,665 in under a year. Catches then stabilized for a few years but shot up again before the war to 46,039, an all-time record of whales caught with forty-one factory ships. After the war they pick up again, but not to such heady levels. They peaked one last time at 38,810 whales in 1960–61, after which they steadily declined in line with whale populations: there were simply too few whales to catch. The whalers had to adapt their hunt to patterns of declining stocks. Besides right whales, the species most favored by industrial whaling was the largest one, which yielded the most oil, the blue whale (weighing between 90 and 140 tons). This was also the species most affected by the stern slipway: in 1928–29, 12,743 blue whales were caught in Antarctica alone, and 17,898 were caught the following year. 1930–31 saw a record catch of 29,410 blue whales. This was also a turning point: blue whales were endangered as early as the 1930s. From then on, the whalers would work down the range of species to the next largest one. By 1936 they were catching more fins than blues (14,381 and 14,304, respectively). By 1937 fin populations started showing signs of vulnerability. The next species down the line was the sei whale, whose catch rate overtook the fin rate in 1964. The industry thus threw itself into a destructive cycle where it was catching more whales and actually producing less oil.

Norway's regulatory framework laid the first stones for international regulations. In 1930 the League's Council for the Exploration of the Seas formally endorsed the Sandlefjord Whaling Bureau as an international clearinghouse for all whaling data and encouraged all whaling countries to submit their catch information. In 1931, the first Convention for the Regulation of Whaling was signed in Geneva by twenty-two nations. It comprised the first milestone in international whaling cooperation, and it established some significant precedents, such as the first recognition of aboriginal whaling (Gambell 1997a). However, it fell far short of what was needed to contain the whaling industry (Scarff 1977, 349). It merely reproduced some of the less stringent conservation measures written into the Norwegian Whaling Act. Furthermore, its entry into force was delayed until 1935, and it lacked key players, Japan, the USSR, Chile, Argentina, and Germany, effectively leaving out 30 percent of world whaling (DeSombre 2002, 124). The Convention halted neither the depletion of the whale stocks nor the collapse of the whale oil market, which reached new lows in 1933 (see figure 2.1).

The British–Norwegian Cartel Agreements and the Blue Whale Unit

The first regulatory attempts were defeated by the behavior of the British state. Given the key role the United Kingdom has since consistently played in the IWC (see chapter 10), these point to a "presocial" stage, where the social pressures from other states were still too underdeveloped to rope states into cooperation. As the second largest whaling nation, responsible for 48 percent of whaling catches in the early 1930s (see figure 2.2), the British state's ratification was both a necessary condition of the Convention's entry into force and of effective regulations. Yet it was delayed for another three years. Beyond Antarctica, British whaling was largely unregulated, and having urged the Norwegians to develop their whaling laws, the United Kingdom resisted Norwegian pressures to develop their own. Only when Norway waved the threat to enact their "crew clause" and to withdraw from a lopsided cooperation that significantly benefited British whalers did the United Kingdom finally ratify the Convention in 1934 (after eighteen other nations had done so) and adopt its first Whaling Law. In other words, these bilateral pressures operated as straightforward material threats rather than as peer pressures. Moreover, the law was largely an empty gesture, as it effectively spared British firms—that is, mainly Unilever—from any type of restrictions (in time or volume, for example). It merely required them to obtain a license from the Board of Trade, which was effectively a certificate of seaworthiness (Tonnessen and Johnsen 1982).

The move toward regulations did continue, albeit carried forth by the industry rather than the state and for the practical purpose of rescuing the whale oil prices rather than the moral purpose of rescuing the whales. When whale oil prices hit a record low in 1931, British and Norwegian whalers, who were often already linked financially, formed their own International Whaling Association and up drew a cartel-like agreement, which was repeated in 1937 and 1938. This self-imposed agreement was the first comprehensive catch management system at the international level. It included catch time and size restrictions and bonuses for switching species, from the threatened blue to fins, for example (Tonnessen and Johnsen 1982, 405). It triggered efficiency improvements in that by capping catches for the first time (at two-thirds of their previous levels), it motivated the whalers to get more out of them. However, these cartel agreements dissolved as soon as whale oil prices rose once again.

At this stage, in what was a good instance of deployment of discursive power, the industry effectively dictated the terms of the discourse within which international whaling regulations could develop. This would have long-lasting, negative repercussions for both the whales and the industry. The Blue Whale Unit (BWU) was the unit of measurement devised by the industry to quantify the oil extracted from the catch. The yield from each whale was translated into "how much" of a blue whale it would take to obtain the same quantity of oil. For example, 2 fin whales made up 1 BWU, or 6 sei, 2.5 humpbacks, or about 60 minke whales. Used by the scientists themselves, this unit of measurement entrenched particular knowledge practices whereby whales were known/measured according to how much oil they yielded. It had the practical effect of fueling a "measuring-up" logic among the whalers and of blurring the differences between species that were endangered and those that were not by bringing all the catches down to the same unit and thus making it easier for whalers to shift targets in the absence of the most desired species, that is to say, the largest (the blue whale). The cartel agreements imposed the BWU as the basic unit of measurement of the IWC up until 1974. For example, all national quotas were set in BWUs. Thus the interests of the whaling industry were provided for by the 1946 Whaling Convention, which wrote in the obligation to "take into account the interests of ... the whaling industry" (Article V.2). But there they were also balanced out against "the interests of the nations of the world in safeguarding for future generations the great natural resources represented by the whale stocks" (Preamble). These interests were more fundamentally enshrined by the fact that the industry's particular knowledge frames had been locked into the nascent regulatory discourse in a way that would inherently advantage it and make it harder to limit whaling.

The International Whaling Commission: A Social Field

The multilateralization of whaling after 1936 and the competition with Japan and Germany forced the United Kingdom out of its reluctance, as it undermined the bilateral cartel agreements. This multilateralization of whaling also crystallized the development of social dynamics between these interacting states and thus the emergence of a full-fledged social field. States came together successively in 1937, in 1944, and more decisively in 1946. Against a background of increasing political tensions, the International Agreement for the Regulation of Whaling signed in London in 1937 regrouped all major whaling nations except Japan. It was, however, yet another watered-down version of the Norwegian regulatory framework. Nonetheless, it established the first restrictive measures at the international level, setting the first ban on a species (right whales of the North Atlantic, down to 100) and the first size limitations (on blue whales), as well as whaling seasons. Interstate whaling negotiations continued throughout the war, yielding a second multilateral conference in London in 1944. Although it never entered into force, it represented the first time whaling countries accepted the capping of catches and the obligation for their whaling industry to report to the Whaling Bureau. Most importantly, it maintained the creation of an international whaling regime high on the agenda for postwar international cooperation.

All the provisions arrived at in the previous agreements were brought together under the new regulatory framework laid out by the 1946 International Convention for the Regulation of Whaling (ICRW), which yielded, three years later, the IWC, the international institution still in place today. The original IWC is known as the "club of whalers." It brought twelve states together to cooperate around whaling regulations.[10] It staked out a new space of social interactions for those states who whaled and, just as importantly, I argue, who wanted be seen by the other states as "whaling nations." The IWC annual meetings offered an international platform upon which states could position themselves, and be recognized by others, as whaling nations. Yet it is interesting to observe how the society of *whaling* states thus taking shape deliberately set itself apart from the broader society of states that was being built in the aftermath of the Second World War. The convention drafters (and notably the American cetologist Remington Kellogg) conceived of the IWC as forming part of the broader cooperative framework of the UN and provided for this possibility within the ICRW itself (Article III.6). There seemed to be no practical reason not to do so, since one of the UN specialized agencies, the FAO, was being specifically established for the purpose of cooperating internationally around the management of natural resources. Yet the IWC member states voted against this, thereby maintaining the IWC as an exclusive club of whalers. That the IWC was about something more than the most effective way to regulate whaling and manage whale stocks appears to be confirmed by the fact that it took another two decades before it was endowed with any real teeth, namely, national quotas, which constituted the only measure that would contain whaling at all. Tabled for discussion since the early 1950s, these were systematically voted down until 1963, including by some of states most inclined to cooperation, such as the United States, who had spearheaded the entire cooperative process, or Norway, who nonetheless first opposed national quotas (before championing them in the 1950s; Tonnessen and Johnsen 1982, 490).

According to its Convention text, membership to the IWC is open to any "contracting government," so long as it pays its fees.[11] In effect, the creation of a multilateral whaling regime enabled those whaling nations who had lost their preponderance in the trade to uphold their place in the international politics of whaling by providing an international stage upon which they could continue to talk the whaling talk, as it were, and maintain an equal say in whaling matters with those new whaling nations that were outwhaling them. It allowed them to punch above their whaling weight. Previously, and so long as Western states had dominated the trade, a state's capacity to establish the terms of international whaling politics was a function of how much it actually whaled. A state's discursive "capacity to define or be defined" in international whaling politics was relatively aligned with their "material capability," to borrow terms from Jutta Weldes and Diane Saco (1996, 372). These two capacities were disjoined by the creation of a multilateral whaling regime, such that the United Kingdom, which held 5 percent of the world's share of whaling in 1963, still had as much say in whaling matters (and notably on the question of quotas) as Japan, which caught 41 percent of the whales (see figure 2.2). By sharing out the vote equally among all member states, regardless of how much (or even if) they whaled, the IWC effectively preserved the Western states' position in international whaling politics in the face of rising Japanese and Soviet whaling. To those states whose whaling was on the demise or almost defunct, the IWC offered a platform for projecting themselves anew as whaling nations as the case of the Netherlands has shown (see chapter 2). And for those states (such as Australia and New Zealand) who were relatively new to whaling, it was the place to assert themselves as whaling nations. These were some of the gains to be had for states that played the cooperative game.

Moreover, this formal voting equality between the contracting governments did not mean that the power to define the terms of international cooperation was equally shared among all states—that is, the power to set the cooperative agenda and to decide who qualifies as properly cooperating or not. Rather, it meant that, in the new whaling regime, symbolic power was increasingly disconnected from the material ability to whale, and reconstituted around other, discursive and social mechanisms that would leave it more concentrated in some positions within this cooperative playing field than in others. One of these key positions was that of the United States, who had carved out a place for itself as the regime's "benign hegemon" that stood at odds with its rapidly waning whaling capabilities.[17] It had drafted the Convention text and convened the 1946 conference, thereby effectively shifting the epicenter of international whaling politics from London to Washington. By becoming the Convention's repository, it had also established itself as the club's gatekeeper.[13] This still bears repercussions on the cooperative process today. For example in 2003, Belize, who, having paid its fees, sent its Commissioner to partake in the IWC annual meeting in Berlin, was not accepted as a member because the U.S. State Department "had not received" its instrument of adherence. The Belize Commissioner confirmed to me that the instrument of adherence had been sent from Belize to the United States by registered express post on May 23, with the conference scheduled to begin on June 16. Belize was declared a member on the second day of the Convention meeting, once the State Department "had found" the instrument and once the most important anti-whaling vote at that year's annual conference had been cast, Belize having indicated its intentions to vote with the whalers (interview with Ismael Garcia). This institutional role has effectively conferred upon the United States additional discretionary powers in the international politics of whaling.

Nor were all states equally free to join, despite the open membership rules. On the ground, membership to the club of whalers was constituted through specific processes of inclusion/exclusion, and it triggered specific dynamics of social belonging. Most strikingly, it was initially set up by excluding the second and third most important whaling nations, Germany and Japan. The way in which Japan was reintegrated is telling: Japan was brought back to the negotiating table by United States in 1951, despite vigorous protests from its Allies (Tonnessen and Johnsen 1982). What was being established in the process was who could be considered as a "good," cooperating nation, worthy of belonging to this club of whalers, but *outside* the cooperative space that had just been created and without recourse to its formal procedures. Moreover, this contrasts starkly with another, more recent, boundary-defining moment, where the formal voting rules were indeed wielded, only this time to exclude: in 2001, in what has now turned into an anti-whaling club (see chapter 7), Iceland was cast out of the IWC by a majority vote, despite having paid its fees and no provisions enabling this in the IWC membership rules.[14] Hence the ability to determine who is to be considered a good, cooperating IWC member is still today not equally shared among all (formally equal) states.

As for Japan, its reintegration into the IWC after the war, at a time when it was still excluded from the UN (until 1956), was seen as a form of reinstatement among the society of peaceful, cooperating nations.

Membership in the whalers' club was a chance to correct a war-tarnished image on the international scene. This desire to belong, to be seen as a good, law-abiding state that has its place among the civilized nations, is key to understanding how Japan has positioned itself in relation to the IWC over the years (although this has now started to change as we shall see in chapter 11). Japan is one of the few whaling states that has never resigned from the IWC, thereby constantly maintaining its whaling under the scrutiny of the other member states. At every IWC annual meeting, Japan has gone to great lengths to demonstrate how its whaling activities observe the collective rules of the IWC. There has in fact been little material incentive for Japan to remain in the IWC: it would be far less costly, both in terms of shielding itself from negative publicity and from restrictions upon its whaling, to opt out of the IWC—as other states, such as Canada, have done without attracting any of the bad press that Japan receives *while* dwelling within the IWC. As for the other initially excluded nation, Germany, its reinstatement within the IWC also served similar image-polishing purposes: the 1982 moratorium vote provided West Germany with the chance to mark its belonging to the society of good, Western—this time—*anti*-whaling states, and it promptly joined the IWC. The desire to belong, to be recognized as a good, cooperating, whaling/anti-whaling nation (depending on the period), is perhaps more acute in the two countries whose exclusion served to set the outer limits of the society of whaling states than in other nations. They are helpful here for illustrating some more advantages to be earned for states that joined the cooperative game.

A few more features of the IWC reveal the ways in which, structurally, the negotiating playing field was laid out unevenly from the onset. Although the ICRW's Article II proclaimed the formal equality of all contracting governments written into Article II, by Article X some were made to count more than others. It reintroduced an implicit hierarchy among these sovereign contracting states in stipulating that the Convention's entry into force requires ratification by at least six states "that shall include the Governments of the Netherlands, Norway, the Union of Soviet Socialist Republic, the United Kingdom, and the United States." Furthermore, from the onset the dice was loaded in favor of some countries by the choice of the languages that were to form the basis of a common understanding. The nomenclature of whale species in the Appendix listed the whale species in their scientific names (in both Greek and

Latin), followed by their equivalents in English, French, Dutch, Russian, Spanish, and Scandinavian. Given the significance of this document as staking out the discursive field for international whaling cooperation, the choice of languages appears surprising. As Tonnessen and Johnsen (1982, 501) point out, the choice of Scandinavian instead of Norwegian was perceived as a slight to Norway and a misrecognition of its role in the emergence of international whaling cooperation. The absence of German and Japanese is also noteworthy, especially considering that Japanese was one of the few languages relevant to whaling that shared neither common linguistic (Greek or Latin) roots nor a common alphabet. It denied the historical role of these two countries in whaling and inscribed their exclusion as deliberate and long-term, rather than circumstantial. Furthermore, the Convention text, "done in the English language" (Article XI), was one of the few international conventions to be monolingual. This endowed some state representatives with more locutionary means than others. This locutionary differential lies at the heart of what Bourdieu identified as "symbolic domination," a type of power that is never explicitly branded *as* power—that functions in fact on its ability to remain unrecognized as domination (Bourdieu 2002b, 173–176). Bourdieu (1983) shows how in high-level social rituals and official ceremonies, to which the IWC annual meetings would belong, the official language becomes the locus of deployment of symbolic power, as it reenacts power differentials between those who "naturally" wield it well and those who are not so fluent. In my experience, this symbolic advantage built into the negotiating situation itself is still tangible to the observer sitting through plenary sessions of the IWC today: the delay caused by simultaneous translation into English, from say, Japanese, tends to dampen the argument and lose the audience's attention. In a context where the press plays a key part in the international politics of whaling, those commissioners who argue directly in English undoubtedly command that added power of persuasion that can make a difference to the ear of the journalists in the audience looking for ways to add color to these otherwise tedious and long, drawn-out state negotiations that they are to convey to the wider public. Furthermore, until 2003, non-English-speaking nations were to cater for their own translation, an added cost that effectively restricted the participation of some developing countries.[15] This linguistic hegemony, I argue, is a condition of possibility for the rise of the dominant anti-whaling discourse, which developed primarily in the English language.

Conclusion: The Dynamics of State Belonging

This chapter studied the emergence of an international regulatory discourse that posited states—no longer the whaling industry, nor the scientists, nor (yet) NGOs—as subjects of international whaling cooperation. It established the language of international cooperation, mirrored in this book, that enables us to say "Norway proposed this" or "Japan objected to that." This new international whaling discourse was significant because it reformulated sovereignty in a way that was crucial to the emergence of global environmental governance. It undid the well-entrenched link between sovereignty and the appropriation of natural resources (Kuehls 1996) and rearticulated it in relation to communal management. Methodologically, the chapter proposed a sociological approach for the study of international regimes. By foregrounding the notion of "common interests," this type of analysis better captures why and how states take part in collective institutions. States are brought to collective behavior by their own interests, specifically that part of their interests that they share in common with other states and that have to do with the existence of this social system that defines them. States want to be part of a society of states because it defines them; it does something for their identity.

The "club of whalers" sharpened two specific images for the states who joined it: it established states as good, cooperating sovereign states, on the one hand, and as whaling nations on the other. Thus the sociological emphasis on interests also draws out the *relational* dimension of identity and that interstate relations constitute an important site of state identity construction. In analyzing the emergence of the international whaling regime, the chapter examined different sites where the whaling identity took shape: first, the bilateral relations between Norway and the United Kingdom, which were key to developing the first regulatory framework and the British whaling identity; second, the conjunction of inward- and outward-orientated processes that constituted the Norwegian whaling identity, and finally the IWC as a site of social interactions. The latter offered a stage for performing states' whaling identities and obtaining recognition by other states, and it was constituted through specific dynamics of inclusion and exclusion.

Lastly, the focus on communal interests sheds light on why a successful collective management of whale stocks was impossible within the original IWC. For even though the "common interests" of "future generations" and of "the nations of the world" had been provided for in the

ICRW, the actual common interests at work here were regulated by a social norm that was fundamentally exploitative. At this particular point in time, states' common interest lay in upholding a tightly closed club of whalers, not a club of whale savers. In other words, the conservation objectives written into the Convention text were bound to be defeated by the particular social dynamics that took hold within this society of states. Hence the way common interests take shape within specific social dynamics, and around a particular *nomos*, is a better indicator of how states will tend to behave, and how such behavior will change over time, than the legal provisions they sign up to, especially given the unenforceable nature of such provisions in the "anarchical society" (Bull 1977) that constitutes the state system.

The main conclusion of this chapter is that states' drive to belong to this social field has consistently been the determining factor of state behavior with regard to whaling. Whether it was a whaling or an anti-whaling society of states—in other words, where the *nomos* lies—is only secondary. What matters most is belonging to it; many states have readily belonged to both. From there it follows that the way in which the *nomos* lies has to be examined in order to grasp how a particular society of states is normatively inclined and, thus, how states will behave. The rest of the book will consider how the *nomos* underlying international whaling cooperation was fundamentally altered by the rise of the anti-whaling discourse.

II

Producing the Anti-Whaling Order

5

The Making of a Dominant Global Discourse

In the late 1950s killer whales in the North Atlantic were favorite practice targets for the U.S. Air Force. Activist David Day (1992, 18) recalls how *Time* magazine "was able to print, without disapproval or concern":

> Killer whales. Savage sea cannibals up to 30 feet long with teeth like bayonets ... the Icelandic government appealed to the U.S. Navy, which has thousands of men stationed at a lonely NATO airbase on the subarctic island. Seventy-nine bored GIs responded with enthusiasm. Armed with rifles and machine guns, one posse of Americans climbed into four small boats and in one morning wiped out a pack of 100 killers....
>
> First the killers were rounded up in tight formation with concentrated machine-gun fire, then moved out again one by one, for the final blast which would kill them.... As one was wounded, the others would set upon it and tear it to pieces with their jagged teeth (quoted in Day 1992, 18).

Whereas *Time* did not receive "even one" dissenting letter at the time, a little over a decade later, such actions would have been made the object of federal prosecution. The "killer whale" in the story would have featured as an "orca" or even the "panda of the sea," evoking the cute cuddly toys slowly cropping up around children's bedrooms, and these GI training methods would have roused general public outcry. For Day (1992), the turning point occurred around 1964, the year the first killer whale—aptly baptized "Moby Doll"—was brought to a leisure park aquarium. In recounting his own activist experience, David Day is also capturing the moment when a new anti-whaling order began to subsume the prior whaling order. Day (1992, 25) is "amused" at the extent to which perceptions about the whales were so rapidly and so completely reversed. This shift that he takes for granted as "natural" or "normal" should instead be considered, I suggest, as the operation of discursive power. Having examined the whaling order, the remainder of the book now turns to the examination of the automatic associations that are still in place today whereby any evocation of "whales" triggers the sense that they are "endangered" and that killing them is "wrong." The three chapters regrouped in this part II examine three different facets of the production of this "common-sense opinion" or *doxa* on whales and whaling, before part III analyzes the factors enabling its reproduction.

Normative Change in International Relations

This shift from a whaling to an anti-whaling order encapsulates a key question that draws together two major research programs in the study of international relations, namely, the issue of normative change. For scholars of global environmental politics, on the one hand, understanding how—and at what point—ways of relating to and valuing a natural resource begin to shift is crucial to capturing how the policies to manage them can evolve (Hurrel and Kingsbury 1992; Elliot 1994; Litfin 1994, 1998; Hajer 1995; Vogler and Imber 1996; Dryzek 1997; Benton and Short 1999, 2000; Harris 2000; Paterson 2001; Mitchell 2006; Princen, Maniantes, and Conca 2002; Barry and Eckersley 2005; Hannesson 2006). On the other hand, the issue of normative change is central to the analysis of nonstate actors and their ability to impact on international politics in other important areas such as human rights (Boli and Thomas 1999, Keck and Sikkink 1998, Finnemore and Sikkink 1998, Josselin and Wallace 2001, Klotz 2002, Joachim 2003). The overall argument of this book is that this type of normative change has to be accounted for by analyzing the rise to prominence of a discourse—or how a discourse becomes dominant. The purpose of this chapter is to understand how a discourse, coined by activists and deployed from the fringes of politics, succeeded in imposing itself as the dominant frames around whales and whaling—the way in which most of us around the globe still think of and talk about whales today.

This, in turn, conjures up tangentially a third strand of international relations literature. Unlike most discourses analyzed in the field of critical security studies, which has tended to focus on official, top-down discourses (Campbell 1992, Huysmans 1998, Weldes and Saco 1996, Weldes 1999, Weldes et al. 1999, Diez 2001, Waever 2004, Bially Mattern 2005, Hansen 2006), whaling presents a rare case where the dominant discourse spread throughout the international system from the bottom up. The last three chapters have shown that, in the early 1960s, the anti-whaling discourse was running up against a well-established dominant discourse that framed whales as a resource to be exploited as a matter of national security. In this case, the new discourse was not dominant because it was spoken by state officials. Rather, it imposed itself to the officials of certain countries as it evolved into a dominant discourse. Subsequently, a significant difference with that field is that the discursive material under examination in the remainder of this book comprises not the speeches of heads of states or foreign policy officials but indeed every day discourses spoken on the ground, at the bottom of the international system, as it were.

The main concern of this chapter is thus how an alternative "discourse of resistance," to return to Foucault's terminology, was able to develop into a global dominant discourse. However, to rest the case for shifting the analytical focus from "norms" to "discourse," and thus for envisaging discourse not merely as a surface reflector of underlying normative changes but as a factor of change in and of itself, I begin by identifying some shortcomings in the ways in which normative change has traditionally been approached in international relations.

Appraising the Power of Discourse: Not Quite "Knowledge" nor "Principled Beliefs"

A normative change points to a fundamental reorganization of the structure of meanings and values—what I have termed, following Bourdieu, the *nomos*—underpinning practices in a specific issue-area of international relations. International relations scholars were quick and correct to locate the source of much change in nonstate actors, specifically NGOs. In fact, this particular field of research has greatly contributed to moving the discipline's focus beyond the state (Lipschutz 1992, Lipschutz and Conca 1993) in order to encompass the dynamics of a nascent global civil society (Shaw 1994, Sklair 1995, Wapner 1996) or even a "world culture" (Boli and Thomas 1999). However, perhaps because these imply situations where the underlying *structures* are transformed, international relations scholars have tended to rush perhaps too precipitately to agency to explain it, thereby maintaining intact the duality of structure and agency in a way that obscures the more complex processes that mediate between them and the ways in which they mutually constitute each other within a dynamic process (Giddens 1979, 1990,

1991). In these approaches, change is brought about by actors, who are themselves driven either by powerful "principled beliefs" or "knowledge," to draw on Thomas Risse's (2006) distinction, which is useful for reading through the literature on nonstate actors. These two types of agency, in turn, have yielded their sets of tools for unpacking the workings of nonstate actors: "principled beliefs" have bred various forms of agency associated either with "pressure groups" (Willets 1982); "transnational activists" (Klotz 2002, Josselin and Wallace 2001); "moral" (Nadelman 1990), "norm," (Finnemore and Sikkink 1998) or "organizational" (Joachim 2003) entrepreneurs; or "transnational advocacy coalitions" (Keck and Sikkink 1998). "Knowledge," on the other hand, has shaped the agency wielded either by "epistemic communities" (Haas 1992, 2004) or "knowledge-brokers" (Litfin 1994). Both of these lenses have been applied to the shift from whaling to anti-whaling, which has been accounted for either by a growing global realization that killing whales was "wrong" (Nadelman 1990, Stoett 1997) or by the effect on the actors involved of cumulated knowledge of the extent to which whale populations were depleted (Peterson 1992, Mitchell 1998). In examining the precise role played by scientific knowledge in the discourses about whales and whaling, the following chapter engages with the latter line of explanation; this chapter takes issue with the former.

Both perspectives suffer from an excessive agentcentrism, from which two major problems stem. First, such accounts tend to miss the more gradual, tectonic-like transformations that are located within the structures themselves—put simply, the effects of old structures worn down. Hence the importance of the sort of detailed analysis of the deterioration of the structural *material* conditions for whaling that preceded in part I. Second, and as a result—since discourses comprise structures too—they remain blinkered to the power of discourse. Discourse is not absent from these analyses; indeed it features amply, whether conceived either as the power to persuade with one's "principled beliefs"—the power of persuasion—or as the power of "knowledge," to replay Risse's distinction. However, it is never a primary mechanism of change, simply because it tends to be conceived as *a product*, instrument, or capability of the agent (for a good example see Holzscheiter 2005) rather than *producing* the latter. By this I do not claim that agents, as physical objects, are called into existence by a discourse. Rather, I claim that the subject-positions agents hold, how they hold themselves in the world, the objects they relate to, and the projects they can conceive of engaging in are "produced" discursively. In this way discourses produce subjects/agents in a very real but not physical sense.

As a result, these approaches fail to render the fundamental reconceptualization of power underlying the turn to discourse (Deleuze 1986). As Karen Litfin (1994, 49) has aptly underlined, in such accounts discursive power *belongs* to preconstituted agents; it does not *constitute* them. Even the constructivist literature, concerned specifically with constitution of the identities of the social actors, tends to revert to the old model of power as property or capability (see, for example, Wendt 1999, 224). Perhaps this owes to the difficulty of carrying the productive power model through to empirical analyses, a problem that surfaces in Litfin's own work, where it also tends to fade out of sight. However, this fallacy also stems at a more fundamental level from the conflation of *power* and *agency*, in the basic Dhalien sense that the agent is powerful because it is able to effect change (Clegg 1989). Thus separating power and agency is a necessary first step to appraising discursive power and for grasping the ways in which it constitutes new actors in the international system and transforms existing ones. It is a necessary step to taking discourse seriously, that is, not merely as the actors' appendage, but as a key factor of change.

Chapter Outline and Some Discursive Concepts

The making of the dominant anti-whaling discourse occurred in three stages, where each stage reinforced the previous one in a snowballing effect. First, environmental activists reframed perceptions and understanding by producing a new discourse on whales and whaling. However, this new discourse was not in itself sufficient to evacuate a preexisting dominant global discourse. The activist discourse became dominant because it connected two other preexisting dominant discourses: the Cold War discourse on capitalism and democracy and a budding global environmental discourse. Second, once dominant, the anti-whaling discourse, in turn, *produced* NGOs. It created new subject-positions that established them *as* actors in the international system. For the anti-whaling discourse both created a new context in which NGOs could make a difference in the international system and it determined the means by which they could act within it. Thus it both produced *and* empowered them in very specific ways. Third, the dominant anti-whaling discourse created new subject-positions in the society of states that "interpellated" existing subjects of international relations—states—in specific ways and rearticulated their identities from whaling to anti-whaling states.

This snowballing process that produced the dominant anti-whaling discourse is analyzed over the three chapters comprising part II of this book. This chapter focuses mainly on the first of these three stages. It begins by examining how activists crafted the new *story-line* on whales and whaling. It then analyzes how the anti-whaling discourse operated as a *nodal point* linking together two preexisting dominant discourses or *metanarratives* and rose to prominence as a result. The second part of the chapter analyzes what exactly the discourse *did*. First, it yielded new and improbable *discourse coalitions* around the issue both at the international level and in key states, such as the United States. Second, it created a new grammar for a globalizing environmental activism. These two dimensions are considered successively.

Two terms are key to understanding the functioning of a dominant discourse, the twin processes of *articulation* and *interpellation*. I draw here on the work of Jutta Weldes (1999), who developed this two-pronged focus in order to operationalize the insights of discourse theory in the analysis of powerful discourses in international relations (see also Weldes and Saco 1996). *Articulation* captures the operation of common sense. It draws out the ways in which a statement implicitly triggers sets of preexisting associations, for a discourse works as a broad web of meanings, such that, when proffered, the terms "summon" one another implicitly and automatically. The chain of connectivity is central to the functioning of the discourse, for it serves both to hold it together and to perpetuate it. *Interpellation* refers to the ways in which discourses carve out subject-positions that "hail" actors in such a manner that they become the "subject"—the "I"—of that discourse. However, recognition plays a key part in the understanding of "interpellation" evoked here in two ways. First, the subject recognizes the discourse as its own—that is, it relates to, appropriates, and endorses it. Second the subject recognizes *itself* as the subject—the one who says "I" in this discourse. It is therefore an active process. This is a key difference between the discourse approach employed here and the so-called dominant ideology critique, where "interpellation" was first coined, drawing on the work of Louis Althusser (Laclau and Mouffe 1985; Howarth 1995, 2000). Indeed, at the other extreme of the transnational actors literature's excessive focus on agency, the latter tended to evacuate it altogether. Rather, in the wake of Laclau and Mouffe's (1985) reworking of the concept, discourses are analyzed as articulatory practices by which meanings are produced and reproduced, challenged and then renegotiated.

A Nodal Point Welding Two Prior Metanarratives

Crafting a New Story Line

In order to analyze the making of a dominant global anti-whaling discourse, let us begin by mapping out the story line with which activists rewrote the discourse around whales and whaling. First coined by Maarten Hajer (1995) to analyze the mobilization of discourse in the policy making process, the concept is relevant here since the discourse under consideration is similarly geared toward the realm of action (rather than merely debate) and indeed policy change—and in fact, because, as we shall see in chapter 7, NGOs were effectively using this story-line to enter into the policy-making process itself. *Story-lines*, in Hajer's (1995, 62) words, constitute "Narratives on social reality through which elements from many different domains are combined and that provide actors with a set of symbolic references that suggest a common understanding." Thus *story-lines* are simply a more specific and applied form of *articulation*. Whereas articulation captures the general functioning of a discourse (how different elements hinge together), story-lines highlight the ways in which problematic situations are framed within specific narratives that account for what constitutes "the problem" and how it came about and, *therefore*, what needs to be done about it. In imposing itself this particular "line of facts" frames out other ways of understanding the problem and, consequently, other policy solutions. It thus paves the way for a specific course of action advocated by a particular set of actors, who form a *discourse coalition*. The key function of story-lines is that they bring closure to often highly complex problems and the promise of a clear-cut resolution. Whereas the anti-whaling discourse is analyzed at length in chapter 9, and the various strategies deployed by environmental activists both on the ground and in the policy arena are examined in chapter 7, the purpose here is simply to sketch out the main building blocks of this story-line around which the anti-whaling movement developed. To reiterate an important point, my purpose here is *not* to take issue with validity of the truth claims contained in this story-line or, to put it differently, to dispute whether it offers an accurate interpretation of what happened to the whales or whether they are endangered. Rather, the aim is simply to unpack how the whaling issue was accounted for within a particular perspective or frames that were soon established as *the only way* of understanding the whaling issue.

Rare, remote, and (sometimes) immense, there has been a long line of cultural productions featuring the whale as a mythical creature attributed with all manner of awe-inspiring qualities, not least the *Bible*'s figure of Jonah. In Melville's *Moby-Dick*, the literary masterpiece of the whaling order, this mystery evoked, with the highly ambiguous figure of the white whale, the malicious, indomitable forces of nature, struggling against whom the modern man (not woman) made himself (Melville 1851). In the anti-whaling story-line these very same qualities are mustered to completely opposite effect: in the confrontation between the whalers and the whale, the whalers no longer stand in lieu of humanity at large but rather they are the deceitful, ruthless, indeed evil agents harboring no respect for the beauty and harmony of nature (see, for example, Hunter 1980, Day 1992).[1] Whales, on the other hand, are magnificent, mysterious creatures who, with few predators in their natural habitat, peacefully wallow in blissful ignorance of the greedy voraciousness that preys over them. They are the perfect icon of paradisal innocence or indeed of the state of nature before it was torn apart by the irruption of evil (paradise) or corrupt civilization (the state of nature). Indeed, whales are just like us, or rather how we would like to be—the anti-whaling story-line emphasizes the mammalian characteristics of the whale, as well as other human-like characteristics such as its intelligence or sociability, to draw out our "natural" proximity with the animal and mark the rift with the whalers, a point that will be further developed in chapter 8. Whaling, on the other hand, represents all the excesses of a dysfunctional, modern society—a society that, in its obsession with "growth," knows only to plunder and destroy nature and, eventually, itself. Crafted against a backdrop of growing social unrest and political radicalism on either side of the Atlantic that fomented a "second wave" of environmentalism (Guha 2000), the anti-whaling story-line pinned whaling as the issue that encapsulated the fundamental choice facing us as modern political subjects: to continue on this insidious path of fraying democratic controls and waning political transparency or to reclaim citizen power and face the need for a fundamental social change so as to evolve toward a more sustainable, harmonious relationship with our environment.

Melville's whale and this anti-whaling story-line draw out two completely opposite ways of representing the same animal. They are useful for separating out the real animal "out there," or *signified*, from its representation, or *signifier*: the signifier is what becomes rearticulated into a completely different story-line in the shift from the whaling to the anti-whaling order. This has nothing to do with real whales, who remain far out in the ocean, quite unaffected by this discursive operation. And yet it has very real, practical implications as to whether in practice whales will remained untouched or not.

The use of images was key to pinning these new sets of meaning onto the whales. One of the Greenpeace group's first "image-events," to borrow a term from media analyst Kevin DeLuca (1999), marked the founding moment in the emergence of the new story-line. In the summer of 1974, Greenpeace activists had tracked down a Russian factory ship in the North Pacific. Paul Watson positioned his Zodiac between the harpoon and the whale, while the photographer shot the scene from the other Zodiac. Another camera was filming from the deck of their boat, the *Phyllis Cormac*. The harpooner fired five feet away from their heads. In the foam a sperm whale lay afloat, and fellow activist Robert Hunter jumped into the water and climbed atop the dead animal, to prevent the Russians from reclaiming it. This, too, was caught on film. Reflecting on the effects the images would have, Hunter wrote:

> As a media campaign, the voyage was already a success. No network would be able to resist such footage, just as no wire service would be able to ignore the story. As a newsman, I knew we had achieved our immediate goal. Soon, images would be going out into hundreds of millions of minds around the world, *a completely new set of basic images about whaling*. Instead of small boats and giant whales, giant boats and small whales; instead of courage killing whales, courage saving whales; David had become Goliath, Goliath was now David; if the mythology of Moby Dick and Captain Ahab had dominated human consciousness about Leviathan for over a century, a whole new age was in the making. Nothing less than a historic turning point seemed to have occurred (Hunter 1980, 230, emphasis added).

These images interpellated the viewer into the position of witness to the plight of the whale, a strategy that explicitly drew upon the Quaker religious tradition of "bearing witness" (Hunter 1980; see also Epstein 2003, Rubenstein 1989). The viewer–witness was presented with two options before the whale, incarnated by the harpooner and the activist: to shoot or to save. The images struck beyond any expectation. They were seized upon by the American media, who greeted the crew en mass in San Francisco, and relayed on television channels in Canada, Europe, and even Japan. They were said to have moved even the future hunter–president Jimmy Carter (Ellis 1992), who, in the wake of Richard Nixon, entrenched a long, uninterrupted bipartisan line of American presidents to put their names to the cause. That is, they stepped into this subject-position that had interpellated them. What activists offered the media, the public, and these presidents was not merely a powerful headline but indeed an entire new story-line on the relationship between men and whales, and beyond, on the causes of environmental degradation. This story-line soon became engraved as "the" way of understanding the issue, not least through sheer power of constantly reiterated images that Greenpeace and other environmental groups have to this day continued to feed to the media in an uninterrupted stream.

Although the new story-line proposed by the anti-whaling activists may seems compelling enough from where we stand today, its dramatic imagery does not in itself explain why it could displace the other long-entrenched way of envisaging whaling in the West. The anti-whaling story-line took hold, I argue, because it welded together two preexisting dominant or meta-discourses. Let us consider each in turn.

Metanarrative I: The Cold War Discourse on Capitalism and Democracy

In his excellent book *Image Politics: The New Rhetoric of Environmental Activism*, Kevin DeLuca (1999) analyzed the conditions under which activist discourses have been able to attract favorable coverage from the mainstream U.S. media. It is noteworthy that, coming at it from a very different field (media/communication studies), DeLuca similarly takes issue with the monolithic approach to domination conveyed by the "dominant ideology critique" for the tendency to reduce all cultural productions to the expression of a dominant ideology. He argues instead in favor of a grounded focus on specific discourses (or "framing processes," in his own terminology) that is better able to tease out the various forms of resistance that may be carved out from within the midst of these ideologies. In his study, as in the approach to discourse proposed here, the focus on discourse serves to counteract the evacuation of agency operated by an excessive focus on ideology. It is the means to capture forms of domination that may originate beyond the frames of the dominant ideology.

De Luca compares the consistently negative media coverage of the Earth First! campaign to save old-growth forests in the United States with the positive attention given to Greenpeace's anti-whaling campaign in the media. The Earth First! campaign suffered in its coverage from its subversive, anti-capitalist overtones that made it incompatible with the frames of a corporate media. Greenpeace, by contrast, was able to tap into the Cold War metadiscourse of capitalism and democracy. Targeting a Russian whaling factory was particularly effective. The images of the Greenpeace activists returning to the San Francisco harbor, reminiscent of the return of World War II veterans, served to reaffirm the "heroic Cold War" (DeLuca 1999, 98). The Greenpeace activists were successfully typecast as "rugged individuals, a key mythic character of capitalism and democracy, versus the dehumanized technological juggernaut of Soviet communism" (DeLuca 1999). In DeLuca's analysis Greenpeace thus succeeded in subverting the media's own frames to convey its broader critique of industrialism, anthropocentrism, and progress directed against those very same corporate structures in which the media is embedded.

Moreover, the way in which Greenpeace settled upon the whaling issue, while not the focus of DeLuca's analysis, is significant here. Greenpeace was originally founded in 1971 as an anti-nuclear protest movement against the backdrop of the American involvement in Vietnam. It was initially called the Don't Make A Wave Committee (Hunter 1980) and *Greenpeace* was the name of the boat purchased to stage its anti-nuclear protests. Soon the same boat that had been used to obstruct atomic testing would be used to impede the slaughter of the whales. As I have argued elsewhere (Epstein 2003), in a Cold War context the group's evolution toward the anti-whaling campaign as of 1974 secured it much surer chances of success, since it pitted it against *another* state rather than their own.

Metanarrative II: Saving the Planet

The entrenched discourse was the discourse on endangered species protection, which was progressively taking shape throughout the 1960s as the first global environmental discourse. Given the range of environmental issues that had been laid on the table during what has come to be known by global environmental historians as the decade of the "environmental revolution" (McCormick 1989; Pearce 1991; see also Brenton 1994) this claim requires some explanation. The 1960s saw the public concern for the global environment soar to new heights. As television sets were rapidly spreading across the developed world (in the early 1960s in the United States, mid-1960s in Europe, and early 1970s in Canada), the ever larger televisual public was confronted with accumulating evidence of environmental damage pouring into their living rooms: a fire in the Windscale nuclear plant in Northern England in 1957, the fallout of radioactive ash from the testing of an American hydrogen bomb near the Marshall Islands in 1954, the first nuclear reactor accident in the United States (Idaho) in 1961, the oil spill from the tanker *Torrey Canyon* off the coast of Britain in 1967. This was thus also the decade of the emergence of a "televisual universe," which former Greenpeace United Kingdom program director Chris Rose (1993, 287) rightly identifies as absolutely central to the emergence of a global environmental movement. This environmental interest rippled through to other forms of media: the new National Aeronautics and Space Administration Lunar Orbiter satellite images of the "blue planet," first published in 1966, would soon become one of the most reprinted images in the world, spun into all manner of metaphors of vulnerability (such as "spaceship earth," "only one earth," etc.). "The environment" rapidly became an important thematic of the new visual popular culture associated with what media analyst Marshall McLuhan famously called "the global village" and linked to the consolidation of global media networks. This found echo in the written culture, too, where a wave of popular literature by the so-called prophets of doom began flooding bookstores as early as the early 1950s.[2] Such was the public sensitivity that a book on the seemingly arcane topic of insecticides—Rachel Carson's 1962 *Silent Spring* immediately shot to the top of the best-seller list. Thus Jacqueline Vaughn Switzer (2004, 18) remarks that the 1969 oil spill off the coast of California "hit a public nerve like never before." Only eight days into his administration, President Richard Nixon was faced with an environmental crisis for which he was totally unprepared. Another commentator from the Brookings Institution captured the *air du temps* in the United States:

> In 1969 and 1970, environmental hazards seemed to be everywhere. If you were watching the evening news on newly popular color television, you would have seen a 400 square mile oil slick, moving toward the beaches of Santa Barbara, California.... Such incidents helped fuel growing public outrage and demands for national action. By the time that ordinary Americans including ten million school children gathered all over the country to celebrate the first Earth Day on

April 22, 1970, the majority of those polled supported federal rules to reduce environmental damage (Brookings Institution 1998).

My point here is not to contest the authenticity of this popular concern. It is simply to underline that, against the backdrop of increasing concern for the state of the planet, "the environment" also operated as a *floating signifier*, with no fixed reference point nor even a clear-cut content but instead temporarily pinned onto the latest disaster erupting upon the news.[3] The distinction drawn here is similar to that with the whales earlier on, namely, between the environment "out there"—immensely complex and still little understood physical phenomena—and "the environment" as the signifier mobilized (whether in written, spoken, or visual form) into political and media discourses (see also Rubenstein 1989). An important focus for discursive approaches are those constantly reoccurring key terms of the political debates—such as "the nation," "the people," "freedom," etc. (Howarth 1995)—whose meanings are not given but rather fixed, and whose fixing sets the stakes for the main political struggle. They mark the key sites of contestation. Crisis or change occurs when they become "unhinged" from their preexisting chains of meaning and rearticulated so as to exclude some of those previous associations. They are sometimes referred to, in a different analytical tradition running from Walter Gallie to William Connolly (1993), as "essentially contested concepts," a notion that has been extensively applied to some of the main signifiers of environmental politics, such as "sustainable development" (Luke 1995, Sachs 2000, Connelly 2007).

Yet the study of political discourses has tended to focus mostly on mature, already well-entrenched concepts, with a long history of contestation over their meaning and usage. Less analyzed are the irruptions of new terms that decisively reshape the political debate, such as "the environment" in Western democracies in the 1960s. Here was thus, I suggest, a situation of *under-* rather than overdetermination, a moment of even greater indeterminacy, prior to the fixing of meaning, and thus even to the possibility of its contestation. It is not that "environment" did not have a specific content or a descriptive function when it occurred in television footage nor that there was any dispute as to what did or did not constitute the "environment." But rather it was not clear what exactly "the environment" was, beyond "that thing" that was being threatened by human activity. In this sense it was a signifier still waiting to be settled. Thus, as the focus is less on the struggle to contest the meaning (since it has not yet been fixed), these moments of underdetermination draw out even more starkly the founding insight underlying the discourse focus, namely, that no word inherently "contains" its meaning, that the process of signification is the fixing of a signifier to a signified, and that it always remains both an incomplete and a contingent process. In this context of underdetermination of the signifier "environment," endangered species protection provided the issue that fastened it.

Protecting Endangered Species AS Saving the Planet: The Construction of a Synecdoche[4]

In the endangered species protection discourse in the 1960s, "endangered species" operated as synecdoche for the global environment, such that acting to protect them served to address the broader problem of environmental destruction. I begin by retracing how this synechdoche was constituted, as its trajectory draws out this fundamental contingency underlying processes of meaning fixation. Two moments were key to the making of this synecdoche: first, at the domestic level, endangered species protection became entrenched as the paradigm for environmental policy making in the United States in the 1960s. Second, at the international level, in the early 1970s endangered species protection was established as the first global environmental issue—the issue that began shifting the collective behavior of states toward protecting, and no longer merely exploiting, nature (Epstein 2006; see also Kuehls 1996 for a broader theoretical discussion). However, the construction of this synecdoche was neither automatic nor obvious, on both of these accounts. Historically, American environmentalism had been marked by two, equally strong, competing traditions. On the one hand, *preservationism* was rooted in spiritual and nationalistic notions of wilderness that had yielded national parks, one of the first environmental policy tools (Nash 1973, Runte 1979). However, it coexisted alongside a *conservationist* current that sought to reconcile the use of nature with its protection. The latter was entrenched in a long tradition of natural resource management that went to the core of the very structures of the modern bureaucratic state (McCormick 1989, Murphy 1994, Knobloch 1996). Furthermore, in the United States, while wild *spaces* had been protected, wild*life* had tended to be used: one of the very first federal environmental laws, the 1900 Lacey Act, protected wild species only insofar as they constituted "game," in accordance with "the 19th century conception of wildlife law—the preservation of a food supply" (Vaughn Switzer 2004, 257). In fact, wildlife law had developed in the United States on the basis of the Supreme Court's *restraining* the federal powers' ability to limit the use of wildlife.

Nor was it automatic that endangered species protection should evolve into the first global environmental issue. Born in the metropoles (particularly the United Kingdom) at the turn of the nineteenth century, endangered species protection developed mainly as a colonial ideal carried beyond the national territory through the consolidation of the empires, where in the organization of colonial rule, wildlife conservation was placed under direct control of the colonial state (Mofson 2000, Adams 2001). It was also at the origin of the creation of the first properly global environmental NGO in 1961, the World Wildlife Fund (WWF). However, the budding UN system had initially entrenched a more use-based approach to global environmental governance. First, the first wave of specialized UN agencies (such as the FAO, the World Health Organization, etc.) were established in a postwar world in reconstruction where the focus lay squarely on utilizing natural resources to rebuild severely destroyed economies. Second, using natural resources was also an utmost priority for the large number of newly independent states in the international system entering the race toward development. Thus the principles that drove the emergence of a supranational layer of global bureaucracies of nature management were geared toward the rational, scientific utilization of resources rather than the preservation of particular species or spaces as indicated by the title of the very first UN specialized conference, the 1949 UN Conference on the Conservation and Utilization of Resources (McCormick 1989). Third, various legal and institutional developments furthered a more comprehensive approach to the global environment, based on the rapid developments in ecological sciences, that moved the focus beyond the single-species approach: the Antarctic Treaty system, launched in 1959; the long, drawn-out law of the seas negotiations, finalized by the signature of the United Nations Convention on the Law of the Sea (UNCLOS) in 1982; and the honing of new tools for global environmental governance within the UN system that sought to operationalize the scientific concept of "ecosystem" into international conservation programs—for example, the notion of "biosphere" furthered by the Man and Biosphere Programme that was launched by the United Nations Educational, Scientific and Cultural Organization (UNESCO) in 1968 (Batisse 1982). How, then, in such context, could endangered species protection successfully emerge as the first global environmental discourse?

Stockholm 1972

The 1972 United Nations Conference on the Human Environment carved out the site for the emergence of a new category of agency in international cooperation: protecting the global environment.[5] Called for by the UN General Assembly in 1968, Stockholm was the first UN "theme" (rather than issue-specific) conference, and it marked the recognition by states that environmental damage was a global phenomenon that had to be addressed collectively. Another important political function was to draw developing countries into international discussion of environmental issues (Brenton 1994). Stockholm set key milestones for an institutional framework of global environmental cooperation, creating the United Nations Environment Programme (UNEP) to collect, coordinate, and disseminate information and policy initiatives at the global level. However, while the wheels of international environmental collaboration were set into motion, the issues to be addressed at the conference could not be settled. The "North" came with their own set of issues of particular concern to their publics—such as marine pollution, acid rain, overconsumption of resources, population growth, and indeed endangered species—but (much to the latter's surprise) these were strongly contested by "the South," who sought instead to direct global attention to the problems of poverty and unequal distribution of resources in the international system (Brenton 1994).

Now, fast-forwarding to consider the range of issues actually addressed in the wake of Stockholm, it seems this vague and all-encompassing notion of "doing something about the environment together" that had been placed on the table by states themselves was in practice pinned onto "protecting endangered species." It is not that all the many other issues on the table were evacuated; rather, as Tony Brenton (1994) has shown, the bulk of the policy-making activity after Stockholm reverted to the regional level—with various environmental initiatives around the Mediterranean, the North, and Baltic seas. In fact, the only two non-species-related initiatives that attempted to generate a global consensus—on marine pollution (the 1972 London Dumping Convention and the 1973 International Convention for the Prevention of Pollution from Ships, or MARPOL)—lay dead in the water for another two decades (Brenton 1994, 92–95). The only *global* consensus to emerge in the aftermath of Stockholm concerned endangered species protection. Indeed, the string of international conventions successfully signed over the rest of the decade addressed various dimension of species depletion: the 1971 Ramsar Convention on Wetlands of International Importance Especially as Waterfowl Habitat, the 1973 Convention on International Trade in Endangered Species (CITES), and in 1979 the Convention on the Conservation of Migratory Species of Wild Animals (Bonn Convention). This was all the more surprising in that, at the conference itself, the issue of endangered species had been relatively sidelined, to the extent that Gerardo Bodowski, the director-general of the International Union for the Conservation of Nature (IUCN), whose focus was still very much species oriented, remarked with disappointment that "the theme of wilderness and the need to maintain and enhance diversity was given little attention" (McCormick 1989, 120). What, then, had happened at Stockholm?

By contrast with the general absence of wildlife on the Stockholm agenda, a special case was made for whales by the American delegation. At the conference the United States found itself in an overall ambiguous position (McCormick 1989). One of its staunchest advocates in the early days, it was rapidly pitted against a series of general agreements taking shape during the conference itself. For instance, it opposed many of the initiatives proposed or supported by less developed countries, thereby undermining the conference's main political objectives. It tried to weaken a proposed International Register of Potentially Toxic Chemicals, it abstained from voting on a resolution condemning nuclear weapons testing, and it opposed the expansion of the proposed governing council of the new UN environmental program. On the issue of acid rain pollution that it was exporting to its Northern neighbor (Canada), it displayed little goodwill—although it was not alone to do so (the United Kingdom, for example, took a similar position). The American stance stirred increasingly critical voices—from Sweden, India, China, Iceland, and Tanzania, as well as from activist groups (both American and non-American) outside the conference hall. On another front the United States came under fire for what the Swedish Prime Minister denounced as its "ecoside" in Southeast Asia (Keck and Sikkink 1998, 120). Against this backdrop of criticism, the U.S. delegation won popular acclaim by championing a ten-year moratorium proposal on commercial whaling, which was adopted by an overwhelming consensus—a 52 to 0 vote. The same proposal was carried straight from Stockholm to the opening session of the IWC's annual meeting by the Secretary General of the conference and future UNEP Director General, Maurice Strong, amid loud clamor and as a token of all that had been promised to the public. Strong declared to the IWC:

> It is recommended that Governments agree to strengthen the International Whaling Commission, to increase international research efforts, and as a matter of urgency, to call for an international agreement under the auspices of the International Whaling Commission involving all governments concerned in a ten year moratorium on commercial whaling.

At the IWC meeting the American delegation also sponsored a resolution requesting that the UN Secretary General address all nations of the world, recommending all those engaged in whaling partake in the IWC, in "the spirit of the United Nations Conference on the Human Environment" (IWC 1973, 27). The European Union promptly placed a ban on the trade of whale products that same year (Stoett 1997). At this particular juncture, the whale became the emblem of a society of states that was taking a green turn.

Endangered Species Protection: From U.S. to International Law

Let us return to the two key moments in the formation of the synecdoche. Beyond the diplomatic effect at the conference itself, the American proposal is best understood in light of domestic developments, both short-term political developments and long-term institutional evolution. The second half of the 1960s witnessed the swift erection of a national framework of environmental laws. Hitherto disparate initiatives were progressively coordinated into the 1969 National Environmental Policy Act, which established both the country's first federal environmental policy and agency (the Environmental Protection Agency in 1970). Yet, as Donald Worster (1977, 256) points out, the broader conceptual backdrop was one where, in the United States, the thinking on conservation had shifted decisively away from the "progressive ideology of utilitarianism" (which he sees ending in the 1920s) toward "preservationist policies," contrary to the developments taking place at the global level. In this context, animals played a prominent part in the emergence of U.S. environmental law. The first protective law was the 1958 Humane Slaughter Act, followed by the Animal Welfare Act in 1966, which met with significant public interest and prompted a wave of stories about abused domestic animals in the media.[6] After pets, wild animals and their habitat soon captured the legislators' attention, with a first Wilderness Act passed in 1964. This attention to wildlife was further refined into a concern for endangered species. In 1966 Congress established endangered species protection as a federal matter with the Endangered Species Preservation Act, which requested a list to be drawn of all threatened species on U.S. soil. Thus, if the "deluge of environmental laws" in the late 1960s was, according to one observer, "the catalyst in bringing ecological values to the forefront of American life," then protecting species deemed endangered was entrenched as the American way of endorsing these values (Brookings Institution 1998). Endangered species protection became the hallmark of American environmental policies. The nationwide whaling ban in 1968 was one of the first applications of the new law.[7] The United States had moved suddenly from a complete absence of whaling legislation to a total ban, and from there, efforts were directed at spreading this ban internationally.

In 1969 Congress extended this list-based protection measure to international species with the Endangered Species Conservation Act, which drew up a worldwide list of wildlife in danger of extinction and banned their import. Three additional defining features of global American environmental policies took root at this juncture: an international outlook, the principle of an extensive listing of endangered species and the use of economic measures as enforcement mechanisms.[8] In 1970 the eight largest whales were listed on the Endangered Species list. This significantly impacted the whale trade, as the United States accounted for about one-fifth of the entire world market for whale products (excluding meat; Scarff 1977). Not all animals, however, were equal in the new endangered species protection regime; some were singled out for additional protection. In 1972, despite many marine mammals *not* featuring on the Endangered Species list, the Marine Mammal Protection Act (MMPA) established a permanent moratorium on killing all marine mammals, barring aboriginal exceptions, in the name of what the U.S. legislators perceived, in their words, as "an international consensus" forming around these creatures (U.S.C. 1972a). In fact, the whole language of the MMPA is an illustration of the new prevalence of the endangered species protection discourse. For example, its opening "Declaration of policy" reads, "Marine mammals have proven themselves to be resources of great international significance, aesthetic and recreational as well as economic" (U.S.C. 1972b).

The Act created the Marine Mammal Commission "to safeguard the interests of these creatures"; it instituted trade embargoes against other countries carrying out activities in a manner harmful to marine mammals. Furthermore, it entreated the secretary of commerce to "initiate the amendment of any existing treaty," including an international treaty, "to make such treaty consistent with the purposes of this Act" (U.S.C. 1972b). At that point, marine mammals, and whales in particular, became the flagship species with which the United States took the lead in the global protection of endangered species (DeSombre 2000, 70).

In 1973 the Endangered Species Act (ESA) was strengthened and extended, notably to cover plants and invertebrates. That same year CITES was signed in Washington, effectively a prolongation of U.S. protective policies onto the international level (DeSombre 2000, M'Gonigle 1980). It provided an overarching legal framework for extending the endangered species protection discourse at the global level. CITES reproduced the ESA's appendix system, ranking each species according to levels of threat and tying these to specific protective policies. Listing a species on Appendix I, for example, required states to ban all importation of "any recognizable part or derivative" of the species (CITES Article I(b)ii). By 1977, all blue, humpback, bowhead, right, gray whales, and most fin and sei whale stocks were listed on Appendix I. The American practice of using trade measures as an instrument for environmental protection was thus written into international law. Market mechanisms had replaced direct intervention in orientating the use of wildlife in exporting (or "range") states, who *also* tended to be developing countries. Indeed, CITES has since attracted considerable criticism, notably among developing countries, who increasingly see it as a neocolonial intrusion into their resource management cloaked in environmental concerns (Hutton and Dickson 2000). Launched from the United States in the late 1960s, endangered species protection became the paradigm that orchestrated the emergence of global environmental policy making in the 1970s. Where the global environmental governance agenda has since diversified, endangered species protection has consistently remained the hallmark of U.S. global environmental policies, through to the 1997 Asian-Elephant Conservation Act, which followed on the heels of the 1988 African Elephant Conservation Act (Epstein 2006).

What the Discourse Did

The anti-whaling discourse coalesced state and nonstate actors in a way that effectively ushered NGOs into international environmental cooperation. Moreover, because it appeared to offer a concrete solution to the problem of global environmental degradation, it laid out a program of action for environmental activism. In these two ways, examined here, the anti-whaling discourse produced NGOs as actors in the international system. That is, it both carved out specific subject-positions for them in the international system and defined the means of their agency—the ways in which they could impact upon international cooperation.

Creating New Discourse Coalitions

Besides these long-term evolutions in the legislative framework, the American proposal for a whaling moratorium at Stockholm is also explained by short-term developments in U.S. politics. Indeed, Stockholm was scheduled only a few months ahead of Nixon's running for reelection. With an American whaling industry almost extinct, whale-saving policies provided him with an inexpensive opportunity to pander to a rapidly increasing green vote. Not only had it by then become one of the largest electoral blocs in the United States but it was particularly important to the Republican candidate for winning over his traditionally Democratic home state of California, which was also the starting point for the save-the-whale crusade. In his analysis of the political mobilization of metaphors, Jonathan Simon (2007) has aptly shown how, in a climate of increasing turbulence and growing tensions between the anti-war movement and the institution of the presidency in the late 1960s, the Nixon administration deployed a broader "divide and conquer" strategy that targeted environmental groups, as the more conservative elements of the social protest movement, in order to shift the pattern of political alliances. In this context, the "War on Drugs" constituted a "metaphorical bridge" between environmentalists and Nixon's center-right majority, as drugs were easily analogized to other toxic chemicals placed in water and airways. So politically salient was the environment at the time that Nixon's Democratic opponent campaigned around the slogan "the environment takes precedence over the economy," a catchphrase that seems difficult to conceive today. It is in this context that Nixon closed the last remaining whaling station in Richmond, California, by executive order in 1971. At this point whales took on the unifying function which they have since continued to play in American politics, both across the two main political parties and across the different branches of government (Epstein 2004).

Moreover, the anti-whaling discourse also served to bind the various green NGOs themselves. The 1960s saw the tremendous expansion of the U.S. environmental movement (Carter 2001), which, partly as a result of this success, was also increasingly fragmented. Many new environmental groups were born of a widening rift with the old conservation movements (Jameson 1996, Adams 2001, Carter 2001, Epstein 2006). Both FoE and Greenpeace, for example, had branched off from the Sierra Club (Hunter 1980, McCormick 1989, Pearce 1991). The endangered whale provided a rallying symbol for these increasingly diverging forms of environmentalism across either side of the Atlantic. It satisfied the older NGOs: after all, this was a traditional animal protection or wildlife issue; it could easily suit the agendas of more traditional groups such as the Sierra Club or the British Royal Society for the Protection of Cruelty against Animals, who to this day maintain a representation at the IWC. Some of these groups used the issue to expand globally, such as the U.S. Animal Welfare Institute, which became the International Fund for Animal Welfare (IFAW). The endangered whale also reconciled conservationist and preservationist standpoints. As for the young, radical groups, whales were also the flagship issue around which NGOs such as Greenpeace and indeed FoE developed themselves (see Pearce 1991 for FoE). Moreover, such was the momentum created around whales that some groups, such as Project Jonah or the Whale and Dolphin Conservation society, formed exclusively around this issue, as we shall see in chapter 7. The anti-whaling discourse thus yielded a wide discourse coalition of environmental NGOs that surfaced at Stockholm. There the whale symbol was entrenched as the unifying emblem of a new global environmental activism that was taking shape. The centrality of this symbol to the identity and development of the global environmental movement may help explain its continued importance for NGOs as a "perseverance effect" today.

At the domestic level, the whale thus brought unity and direction to the environmental movement, and it yielded an unlikely alliance of a Republican president and radical protest groups. At the international level, it allowed the United States to cater to this new obligation that was slowly taking shape around Stockholm for states to cast themselves as "green." A key effect of the Stockholm conference was that it provided NGOs with an entry point to the official interstate polity, a phase that was key to the deployment of the international environmental movement

(Frank et al. 1999). Indeed, the number of NGOs effectively grew by around 30 percent in the subsequent decade (McCormick 1989, 124). Stockholm created a global green platform where state delegations and environmental activists came together for the first time, for a parallel NGO forum was held alongside the governmental one, the Environment Forum, which over four hundred organizations attended (DeSombre 2002, 73) and where the first green NGO daily conference newsletter was launched, *Eco* (the precursor to today's *Earth Negotiation Bulletin*), which remains to this day the official NGO newsletter at the IWC meetings. One of Stockholm's most enduring legacies is that it created these open channels of exchange between state delegations and NGOs that remain key to the functioning of the IWC and many other international environmental organizations. Stockholm marked the moment where environmental groups shifted from being social movement outsiders to legitimate policy advisors. However, here too endangered species protection was the issue that carved out their role. Groups such as the WWF and the IUCN (which comprised a loose alliance of state and nonstate elements) had helped write the CITES treaty. They then joined forces with governments to persuade countries with large markets, such as China and Japan, to adhere (DeSombre 2000, 56). Furthermore, it rapidly generated a new wave of NGOs, such as TRAFFIC, formed in 1975 as a voluntary information-gathering agency to help the CITES secretariat monitor the international commerce of wild species.

Creating a Grammar for a Global Activism

The anti-whaling story-line offered a new syntax for activist *praxis*. That is, it did not simply confer a narrative on the endangered whales but provided a script for doing something about it. Because the anti-whaling discourse was constituted as an *activist* discourse, the whale symbol operated a double synecdochic transfer whereby it stood for the environment as a whole, such that saving the whales became shorthand for saving the globe. We have already seen how "species" were constituted as a synecdoche for the environment by the rise of an endangered species discourse; whales, in turn, were a synecdoche for endangered species in the anti-whaling story-line. This double synecdoche is captured by the slogan with which Joan McIntyre, founder of the anti-whaling NGO Project Jonah, rallied activists from around the world at Stockholm:

> How can we save the environment if we cannot even save the whale?

Here the whale stands for all whales, which stand for all endangered species and the environment as a whole. The structure of the synecdoche is apt for articulating the passage from "green speak," to borrow Harré, Brockmeier, and Mühlhäusler's term (1999), to green *act*. I want here to hold up a discourse lens to what Margaret Keck and Katheryn Sikkink (1998) have identified as NGO's "symbolic politics," in order to illustrate how a discourse approach can further the analysis of activist practice. In their account, activists create "symbolic events," such as, indeed, the Greenpeace media event, to draw attention to a problematic practice. The dramatization of the issue serves to generate "a simple causal line" that redistributes guilt and responsibilities. This simplicity is strategically important, as complex causal chains are "less attractive" for the purpose of organizing political mobilization (Keck and Sikkink 1998, 25). But how exactly is it achieved? The synecdoche, I suggest, provides the syntax for operating this simplification that is so essential to activists for developing a program of action. In the face of complex global environmental issues, the whale synecdoche provided a much needed bridge between the realm of knowledge and the realm of action. The synecdoche is a figure of speech commonly mobilized in practices geared toward effecting some form of immediate and spectacular impact or change, or even a "magical effect." Indeed, ethnologists study the phenomenon of "synechdochical magic," in religious practices for example. More broadly "synechdochism" in ethnology constitutes a set of beliefs or practice in which a part of an object or person is taken as equivalent to the whole, so that anything done to the part is held to impact the whole. It is also one of the most common rhetorical tropes utilized in advertising (Chandler 2007, Jhally 1989).

Conclusion

This chapter has analyzed how the anti-whaling discourse, which first took shape as a "discourse of resistance" against a global whaling discourse, successfully imposed itself as the dominant global discourse on whales and whaling. The anti-whaling discourse functioned as a nodal point that joined two preexisting metadiscourses: a discourse on capitalism and democracy, and a nascent global environmental discourse. This, in turn, was achieved by the double synecdochic move performed by the anti-whaling discourse, whereby "the whale" stood for all endangered species, which stood, in turn, for the endangered planet as a whole.

Thus what made the discourse powerful was not the content of the anti-whaling discourse itself—the new story-line it proposed on whales—but rather a specific set of powerful articulatory practices operating in its midst: the constitution of this synecdoche that connected two prior dominant discourses. The analysis of the emergence of this powerful new discourse draws out the fundamental contingency underlying the operation of discursive hegemony. That is, this specific set of articulatory practices was necessary to constituting this new discourse as a powerful discourse; however, the new set of meanings pinned upon the whale by these articulations were themselves contingent, in that the very same animal could have been perceived completely differently—indeed historically they were. The analysis also highlighted the function of nodal points in political discourse and practice: the relationship between all signifiers and signified, inherently contingent, is fixed through the processes of signification, which remains always a partial and incomplete process. This unfixity and contingency jump to the fore around the key "floating signifiers" structuring the political debate. Hence all political practice aims to establish nodal points to fix the meaning of these signifiers. In this particular instance of environmental activism, this was achieved through articulatory practices that were image- as well as word based.

After analyzing how the anti-whaling discourse rose to power, the chapter examined what it did. The anti-whaling discourse created a vast discourse coalition of anti-whaling state and nonstate actors that formed the basis of the state–NGO relationship that remains pivotal to the dynamics of the IWC today (see chapter 10). As it rose to prominence, the anti-whaling discourse interpellated both states and members of the public around the world into "caring for" and "doing something" about the whales. It wrote out a grammar for global environmental action, both for state and nonstate actors, that effectively blurred the domestic–international divide. More specifically, it created new subject-positions: it offered a new "I" to both individuals and states alike. With regard to the general public, it carved out a new subject-position as good, environmentally minded citizen of the world, as chapters 8 and 9 will further demonstrate. As for states, it carved out new subject-positions as caretakers for the global environment. For a state (whether it had been involved or whaling or not), taking on the plight of the whale became a way of establishing oneself as an "environmental" state, and building its green credentials, both nationally, vis-à-vis this new constituency of environmental subjects that was taking shape, and internationally, vis-à-vis other states who were also attempting to go green (see chapter 10). For these dynamics included a collective dimension, such that "the whale" became the marker of a new, green, society of states. This subject-position, once taken on, effectively rendered the sorts of declarations about resuming whaling proffered by the United Kingdom at the 1966 IWC meeting (see chapter 2) simply inconceivable, both for the United Kingdom itself and the other states it was interacting with in the IWC. The new subject-position into which the United Kingdom was interpellated modified its identity. In other words, what articulated the shift from a whaling to an anti-whaling nation for both the United Kingdom and the United States was not the fact that they were no longer involved in the activity on the ground (since, although not much whaling was occurring in either country by the midst 1960s, they still claimed their positions as whaling states at the IWC annual meetings), but this new subject-position by which they were interpellated and which they fully endorsed.

As for the "new" actors in the international system who had produced the discourse in the first place, namely, NGOs, they too were produced by the discourse's rise to power. The transnational discourse coalitions in which states and NGOs shared the same language on whales and endangered species more generally effectively "hailed" NGOs onto the international stage, on par with states. Whales and endangered species orchestrated the transition from NGOs such as Greenpeace being considered as radical anti-government activists to valid interlocutors in the global environmental policy-making process. The anti-whaling discourse thus carved out specific subject-positions for NGOs within the international system. It produced them in very specific ways that were both enabling, in providing a grammar for a globalizing activism, but also constraining, in that they constrained their range of action to a very limited set of issues. Thus, for example, Greenpeace would not be able to carry this newly constituted international agency over to the issue of nuclear weapons, its founding issue. However, this does not mean that they were relegated to always playing second fiddle to states, as mere policy advisors at the meetings of certain international organizations, such as CITES or the IWC. Once established into these subject-positions, they rapidly moved out of the shadows of states and evolved into full-fledged actors in the international system. They successfully established themselves as the third term articulating the relationship between states and a new "global civil society" that was taking shape (Wapner 1996). They defined new forms of politics cast at the global level: to replay the three other categories in Keck and Sikkink's (1998) typology of activist practice, NGOs became all at once the actors to which states had accountability with regard to environmental policies at large and the actors pressuring states and disseminating information among an increasingly global public. In addition, in the specific case of whaling, as we shall see in chapter 7, NGOs thus were empowered and, from these subject-positions, went one step further than in other issues, in actually establishing themselves as international policy makers, in lieu of states.

6

The Power of Science?

Can science provide a knowledge basis upon which to build a common, rational language that would overcome political differences? Can it, in other words, enable the formation of the consensus necessary to develop successful collective policies for the management of whales and whaling? These were certainly the expectations driving the founding of the IWC, whose Convention text established science as the basis for decision making.[1] This chapter problematizes a set of common expectations (or, to use Bourdieu's term, *doxa*) regarding science and its relationship to policy making that underpin the IWC. The anti whaling discourse, for one, stems at its core from the scientific question of endangeredness—the question implicitly posed to the experts of whether whales are endangered. A corollary notion is that increased knowledge about their levels of endangeredness will "naturally" foster protective attitudes toward whales. What these assume is both the possibility of such definitive knowledge and that science has the power to bring us "the truth" about whales, from which the appropriate course of action will automatically derive, as some functional response to "the truth." At a more fundamental level, these assumptions are steeped in a civilizational faith in the power of modern science as the authoritative discourse on "how the world is" (Aronowitz 1988). As such, it is expected to provide a common language of rationality that will enable consensus building. Lastly and consequently, these common expectations also typically underpin the scholarly analysis of the role of science in policy making and international relations, such as in the epistemic community approach (Haas 1992, 2004) which has been applied to the whaling regime (Peterson 1992, Mitchell 1998).

The whaling regime appears to present all the conditions for science, organized in the form of "epistemic communities," to influence international cooperation. First, it offers a complex, highly technical policy problem, characterized by high levels of uncertainty (Haas 1992, 12). Such uncertainty generates the demand for scientific knowledge among policy makers, which tends to be compounded by crisis or shocks (Haas 1992, 12–14). The IWC too was created against the backdrop of a long, drawn-out crisis, constantly declining whale catches (see chapter 4). This precipitated policy makers' need for science, and for a different kind of science—one that could tell them how many whales were left and how many could be safely caught, not one that studied the make-up of whales, as we saw in chapter 3. Historically, moreover, the establishment of the IWC coincided with an increasing "professionalization" of the international policy-making process (Haas 1992, 10–11). Yet as international relations scholars themselves were quick to recognize (Peterson 1992, 148), whaling stretches the epistemic community thesis, in that more and better organized knowledge did *not* lead to more effective policies. If one is to put such faith into science, the truth about whales was known since 1931. Yet it took another five decades for something to be done about it. More strikingly still, the decade-long effort to reorganize and develop IWC science on the heels of the 1982 whaling moratorium was followed by nothing short of a breakdown in international whaling cooperation as of 1993, as we shall see. However, the point here lies not so much with the analyses of the case as the fact that, when applying the epistemic community approach (Peterson 1992) even with an explicitly discursive twist (Mitchell 1997), scholars tended to reproduce an uncritical faith in science. Because the social context and relations of power in which the production of knowledge is embedded are ignored in this approach (see Epstein 2005, Lidskog and Sundqvist 2002 for a similar critique), the power of science is largely perceived as a benign capability that can enable international cooperation. As Karen Litfin (1994, 186) has aptly critiqued, this conception, in turn, is steeped in "the modernist belief that science transcends politics, that knowledge is divorced from political power" (for a critique of its agentcentrism, see also Checkel 1998, 329). Science is thus seen largely as the locus of consensus building rather than as a strategic site of a struggle for power. This narrow understanding of the power of science has, in turn, obscured from the analysis (Peterson 1992, Mitchell 1998) the wide range of power effects at play in the whaling regime from its inception, as we saw in chapter 4. As a case of failed cooperation, whaling drives home the need for a more critical understanding of the use of science as power (Aronowitz 1988).

Science is thus envisaged here as the authoritative discourse on truth, regulating both what can and should be known within specific discursive orders, which constitute, in turn, particular "regimes of truth"—respectively, that of the whaling order and that of the anti-whaling order. As a result, in a discourse perspective, the question of truth is suspended in order to shift the focus to what Foucault (1982) identified as *truth effects*. In other words, my purpose is not to adjudicate on any of the questions that are addressed to the science, such as "Are whales really endangered or not?" and "Are they intelligent beings?" This would amount to replying *from within* the science. Rather, my purpose is to approach science *from without*, as one among several discourses (albeit a very particular one) that form part of a broader discursive order. Michel Foucault's concept of *episteme* serves to articulate this move. An episteme is, in Foucault's (1972, 191) own words:

> . . . something of a world-view, a slice of history common to all branches of knowledge, which imposes on each one the same norms and postulates a general state of reason, a structure of thought that the men of a particular period cannot escape.

The concept is not used here as practical concept to analyze the case but rather to draw out the epistemological underpinnings of the discourse perspective. It brings historical relativity to the analysis of science by drawing out that science is not one but multiple, that it is the product of an epoch (see also Latour 1986). Although the concept suffers from the Foucauldian tendency to evacuate agency altogether—a point to which I return in the conclusion—it is nonetheless useful for questioning the notion that scientific knowledge gives us a direct take on "the world as it really is." Rather, our ways of knowing the world depend on specific discourses in which these knowledge practices occur. We know the world through certain "tropes" that prevail at particular moments in time and enable certain questions to be asked and not others.

The concept of episteme undermines the equation whereby more knowledge leads to better access to "the world as it really is," and therefore to the development of policies better adapted to it. For it problematizes that correspondence theory of the world underlying this equation and, thus, the literature on science in policy making that builds on it. This is because, in the historical perspective, knowledge categories taken as fixed, timeless, and true today may ultimately appear as outdated constructs, out of many "ways of knowing," belonging to a particular time (our own) rather than the best form of knowledge altogether.[2] Returning to my purpose in this chapter, then, it is thus to analyze the conditions of possibility that allowed for the questions of endangeredness and whale intelligence to be raised in the first place; to observe what forms of knowledge the question, in turn, produced; and to observe the political effect to which statements on whales' endangeredness are mobilized. Thus the relationship between science and activism, on the one hand, and science and policy making, on the other, remains central to this analysis.

The chapter is organized according to the scientific concept or procedure at the core of the IWC's successive management schemes: the BWU, the New Management Procedure (NMP), and the Revised Management Procedure (RMP).[3] Furthermore, because developments in IWC science tended to respond to broader evolutions in resource management at large, these two aspects, science within the IWC and without, are kept under examination through each period. In the age of the BWU, which was the unit of measurement of the whaling order (as we saw in chapter 4), despite the demand for science that founded the IWC, the emergence of an autonomous network of international experts—an epistemic community—remained fraught with difficulties. By contrast, the era of the NMP was marked by an increasingly autonomous science. This autonomy was possible, I argue, not because the science improved or it finally succeeded in convincing policy makers to listen but because of a broader revolution in the discursive order—running with the epistemic community terminology, this would correspond to the "crisis" that intensified policy makers' demand for science. Lastly, the RMP inaugurated an increasingly organized science, a science that was also increasingly capable of addressing the questions that had been posed to it by policy makers. Yet science was also increasingly sidelined within the whaling regime. Thus this period, in which we still find ourselves today, could be characterized by *nondemand* for science on behalf of policy makers. And yet, paradoxically, claims to "scientific truth" continue to remain a key feature of the anti-whaling discourse. I return to this paradox in the conclusion.

The Blue Whale Unit: Science Undermined (1949–1963)

As we saw in chapter 3, the establishment of a discrete scientific body in the new international whaling regime did little at first to autonomize cetology from the national structures of power/knowledge where it first took shape. For all the Convention's provisions regarding the importance of science, the very first scientific body set up in 1949 served merely as a low-profile advisory auxiliary of the Commission. When the IWC's Scientific Committee (SC) was eventually created in 1951, it remained without any funds of its own and played little more than a nominal role in IWC deliberations. Furthermore, the composition of the SC was decided upon by the Commission, since each member nation nominated a national representative, who was sometimes simultaneously employed primarily by the whaling industry (allegedly the controversial Dutch scientist Slijper, for one; Schweder 2000). Thus in the early days of the IWC, scientific and national whaling interests remained deeply entwined. Overall, the SC struggled to establish itself as a proper center for international whale research: little was done to encourage transnational collaboration, and most research continued to be undertaken by individual nations. For example, one of the SC's rare requests for funding in 1955 to set up an international symposium on "Whale Research Problems" to resolve an impasse on fin whale quotas was rejected by the Commission (Schweder 2000, 79). Some collaboration did occur, within ad hoc subcommittee meetings convened on his own initiative by the IWC Chairman (in 1953 and 1954, for example). The IWC was thus a low-profile establishment with neither the authority, the clout, nor indeed the voice that whale experts have since acquired. For example, only four scientists attended the 1949 meeting, five the following year (Schweder 2000). From 1953 to 1959, mean attendance at the SC meetings was twelve scientists. The "scientific establishment" of the IWC during the period consisted of twenty-one men in total, compared with an average of fifty members in the 1970s and over a hundred in the 1980s (Aron 2001). Moreover, the SC's work was initially unpublished and hardly reached beyond the small circle of experts. Indeed, the Commission's first annual reports, inaugurated in 1950, made little room for the science (Donovan 1999). Although the SC did initiate its own separate reports in 1955, these did not include the papers produced by individual scientists until the early 1970s. Furthermore, and in marked contrast with today, these papers were brief, they were not compiled in any accessible IWC record, and few were published elsewhere.[4] Thus, until the 1970s, few channels were exploited through which the scientific knowledge could be disseminated to the policy makers, let alone environmental activists or the wider public.

This lack of autonomy was accentuated by limited organized contact with other relevant scientific bodies—the International Union of Biological Science, the International Council for the Exploration of the Sea, or the Biometrics Society. As a result, the IWC was sidelined from a series of major developments taking hold there. First, in the early twentieth century, cetological research remained modeled on the old natural science paradigm of *the organism*. The latter had emerged out of two centuries of naturalist efforts employed at studying and classifying species and reproduced an essentially static and monadic conception of nature (Worster 1977). As a result, whales were studied as individuals rather than as stocks. Meanwhile, however, the emergence of ecology as a scientific discipline shifted the focus to the study of the relationship between organisms and their environment (Anderson 1981), rendering this paradigm obsolete across the natural sciences at large. Thus, in the late 1930s and 1940s, the work on animal ecology became strongly quantitative with the incorporation of evolutionary dynamics and developments in the new field of population ecology (Adams 2001). These evolutionary notions were, in turn, challenged by a "systems" paradigm spreading through scientific circles throughout the 1960s (Jameson 1996, 229). Here the focus was on synthesis and on the dynamic equilibrium of the system as an interactive whole (Worster 1977). Whereas cetologists were still slicing up individual whales on board the factory ships, the natural sciences at large had thus shifted from a paradigm structured around the organism, or individual species, to one centered on the ecosystem.

Second, the IWC failed to partake in a major restructuring of scientific collaboration on a global scale. The broader postwar question of international collaboration around the management of natural resources prompted the development of what Boli and Thomas (1999) have called a "world science system."[5] One model for international scientific cooperation was the 1957–58 International Geophysical Year, organized under the auspices of the International Council of Scientific Unions. In 1964 the International Biological Programme triggered a vast, global effort to gather data on areas of scientific importance, and notably on the new "environmental" questions, for which a distinct scientific committee was established in 1969 (the Scientific Committee for Problems of the Environment).

Lastly, the IWC was bypassed by progress in fisheries management. Fisheries science had developed since the second half of the nineteenth century in reaction to declining stocks in the North Sea (which had yielded the first International Conference on the Exploration of the Sea in 1899; Adams 2001, 39). In fact, some of the leading ideas in population ecology had been provided by fisheries biology, notably W. F. Thompson's work on the Pacific halibut in the 1920s. Thus, within fisheries management, the principle was becoming entrenched of using science to determine what portion of a stock could be taken without exceeding its net rate of reproduction and thereby reducing its overall size. By the 1930s the first mathematics of a maximum sustainable yield (MSY) had been developed, which William Adams (2001) sees as a forerunner to the concept of "sustainable development." They were to be rapidly improved by a cross-fertilization with the rapidly growing field of statistics.

Within the IWC itself, research was also limited throughout the 1950s by the profile of the scientists. The SC was composed of cetologists, all with good knowledge of whales, but no scientist from the rapidly evolving field of population dynamics. This was reflected in their recommendations for research in the second report to the Commission (1951). The twelve items listed (including food studies, swimming and breathing behavior, embryo and pregnancy studies, and the relative frequencies of sighting single whales and schools of different numbers) were all significant to the biology of whales (as individual organisms) but of little value to population estimates, the piece of knowledge most relevant to the setting of quotas (Aron 2001, 106). This was compounded by the complexities of studying whales, who are inherently difficult to track, as they swim in very deep waters and their migration patterns span the globe. Meanwhile, however, within the FAO, developments in statistics and quantitative methodologies were being combined with emerging computer sciences to give rise to new instruments for understanding the dynamics of fish stock.[6] This shifted fisheries science toward a type of applied knowledge that was more relevant to management questions than most of the naturalist-type research still undertaken in the SC.

Throughout this early phase the SC remained unable to wield the necessary authority to stand up to whaling states and the industry. For the problem of overexploitation was in fact well-known by the scientists, but when they voiced their call for restraint, it was simply not heard. As early as 1951 the SC initiated discussions on an excessive take of fin whales, to which the industry was increasingly turning because the Antarctic blue whales were already decimated.[7] Yet it held back on formulating any recommendation to the Commission for another four years, until the dramatic 1955 increase (from 17,474 catches in 1951 to 26,000 in 1955; Aron 2001, 107). That year scientists Ruud and Laws were able to demonstrate that the stocks of fins could not sustain current harvests. The SC wanted to recommend the curbing of catches at 19,000 animals but believed that "in view of the circumstances, a cut of this magnitude would scarcely be acceptable for the season 1955–56" (quoted in Aron 2001, 107), and accordingly it tuned down its recommendation to 25,000 fins, or 15,000 BWU. That the scientists had to translate their catch recommendations into this unit of measurement (barrels of oil) imposed by the industry also tended to mask the urgency of their message. Throughout the rest of the decade the take of fins remained at this level, despite the SC's repeated calls to reduce it and despite hard evidence that fin whales were also disappearing: the industry had simply been unable to fill their quotas. In 1955, production fell short of 25 percent of the quotas. Even when the quotas were finally reduced from 14,000 BWU to 3,200 BWU in 1967–68, the industry could only produce 87.5 percent of that figure. The scientists were further undermined by the states' hard-nosed resistance to national quotas, which made it hard to formulate constraining recommendations. For example, when in 1959 Norway, at the urging of the scientists, proposed national quotas at an IWC meeting once again, the Netherlands simply resigned, which, in turn, led to Norway's resignation. (Both nations eventually returned in 1962.)

The SC was further weakened by regular episodes of internal dissent. For example, during the "fin whale controversy" spanning the years 1951 to 1959, the Dutch scientist E. J. Slijper systematically undermined any research that tended to prove the need to reduce catches (Schweder 2000). Although it did undermine its internal coherence, the problem of consensus was not, however, the main impediment for the SC, as the reports to the Commission were drafted so as to reflect the majority opinion. What weighed it down most during the decade was a crippling concern with "reasonableness," which contrasts sharply with the boldness of some of IWC scientists in the years to come and reflects the SC's awareness of its marginalized position within the whaling regime. William Aron (2001, 107), an SC member from 1972 to 1977, captured the "ethos of the time":

> After a session in which I was fairly vocal about the data demonstrating the need to lower quotas, a more experienced colleague pulled me aside and told me that,

even though he agreed, the proposed reductions were too severe and the Commission would simply ignore the advice if it were proffered.

The New Management Procedure: Science Unlocked (1963–1974)

The Committee of Four

In view of the impasse on the fin whale controversy, in 1960 the Commission requested the appointment of a separate committee of scientists specifically "taken from countries not engaged in pelagic whaling in the Antarctic" (IWC 1960). The "Committee of Three" inaugurated a new phase in the relationship between science and policy making. The American Douglas Chapman, the New Zealander K. Radway Allen, and the Britisher Sidney Holt were distinguished population specialists, each with considerable experience in providing advice to fisheries management authorities, but none with the whaling issue. They were thus unencumbered by the SC's moeurs. At the height of the postwar peak kill (66,900 kills that year), their charge was to derive an Antarctic catch limit and to recommend conservation measures to render whaling sustainable. They had some measure of institutional autonomy, as their reports did not have to pass through the SC.

The Committee met both separately and as observers at SC meetings. They depended on the latter for data sets and strove to maintain a close liaison between the two groups. This division of labor effectively created an external eye for the SC. This served to immediately unlock debates in the SC—for example the mathematical errors in Slijper's arguments (he was an anatomist) were quickly drawn out—and feed constructive critique back into it. Thus the Three's initial report highlighted necessary changes in IWC science. It proposed ways of improving the "collation and tabulation" of data, called for an increase in technical means to handle the new computations, and placed a new emphasis on field surveys of whale stocks (Aron 2001, 109). The Committee became the "Committee of Four" with the addition of John Gulland, another population dynamicist from the FAO. In 1963 their final report was presented to the IWC annual conference. It was the first rigorous quantitative study of the population dynamics of Antarctic whales. The Committee recommended a complete cessation of blue and humpback whale catches and a fin quota of 7,000 or less. They repeated, this time with insistence, the need to eliminate the BWU.

For the first time, the recommended reductions were unequivocally drastic. This audacity was new, as was indeed the reception of the report by the Commissioners. Because the Committee of Four had been appointed by the Commission (rather than by the SC), it was difficult to ignore their recommendations. Furthermore, the SC, no doubt also emboldened, not only seconded the report but went further in their recommendations, urging notably for the cessation of fin whale takes for eight years. They used the report to argue for a significant reduction of the overall quota to 10,000 BWU, down from the 16,000 BWU 1946 quota (Donovan 1995). These recommendations met with mitigated success. The BWU was continued as "the only practical method that could be administered," and the fin quota remained at 14,000 BWU. Nonetheless, two of the most vulnerable species, the humpback and blue whales, were given total—but long overdue—protection in 1964. The real change was that the scientists had made themselves *heard*.[8]

A new scientific front was slowly forming that would progressively sway the relationship between scientists and policy makers. With this external input, the SC began reorganizing from within. More and more qualified scientists attended the meetings, which gained greatly in quantitative competence. Member nations added scientists with strong mathematical skills. Both Chapman and Allen became members, and soon, respectively, chairs of the SC. Holt and Gulland maintained their presence as observers and advisors. By 1972, thirty full-time cetologists filled the ranks of the SC (Peterson 1992). Its work was completed well ahead of annual meetings, and recommendations were formulated generally by consensus the week before the Commission met. The scientists could begin to face the Commissioners as an organized and homogeneous group.

The "Stockholm Effect"

As the scientific body was consolidating both its technical competence and assertiveness within the IWC, the UN environmental conference held in Stockholm in 1972 further buttressed the position of science from without. Casting the growing public concern for the environment onto whales (see chapter 5) served to draw the IWC out of its confinement and to place it at the forefront of institutional developments in natural resource management. The IWC subsequently acquired a permanent office in Cambridge in 1976, with a cetologist, Ray Gambell, as first full-time Secretary. The IWC had evolved out of the files of a subdepartment in the British ministry of agriculture into a full-fledged institution. In 1978 IWC science was granted its own Research Fund, which was set up apart from the General Fund of the IWC, another important step toward establishing its autonomy.

Yet from the scientists' perspective, Stockholm had been highly ambiguous in its celebration both of science and of whales. Indeed, as detailed in chapter 5, when the draft of the American moratorium proposal was under deliberation, there had been no scientific discussion on the state of whale stocks. To quote one observer at the time:

> That the draft was put to a vote without debate can only be regarded as a political move. No consideration whatsoever was given to the position of the Japanese delegation which had twice expressed its desire during the deliberations to refer the draft to scientists for discussion (Scarff 1977, 367).

Perhaps this explains the mixed response of the IWC's SC to Stockholm's call for a whaling moratorium. The world's attention had indeed goaded research within the IWC. At the 1972 annual meeting, the Commissioners charged the scientists with the task of establishing the factual base for assessing the need for a whaling moratorium. The following year the SC came to the unanimous conclusion that a blanket moratorium could *not* be justified on scientific grounds and recommended instead an expanded research program (Aron 2001, 111). These views were reiterated the next year and were cited by the Commission in voting down the moratorium proposal. This marked the first disjuncture between the SC and the anti-whaling movement. Ironically, it placed the scientists on par with the whaling industry.

The debate nonetheless triggered further improvements in whale management science. Instead of a blanket moratorium, the scientists proposed targeted moratoria on overharvested stocks. The commissioner for Australia—then still an active whaling nation—presented the Commission with a comprehensive reform plan, the NMP, readily received and adopted by the Commission in 1974. It was based on the calculation of an MSY, a management tool developed in the late 1950s as an application of developments in population dynamics. With the MSY, resource management was refined as the principle of maintaining populations at optimal sizes so as to yield the largest harvest indefinitely.[9] The BWU was abandoned, indicating the shift from an economic (the potential oil yield) to an ecological (whale stock characteristics) criterion. The NMP took whale management largely out of the hands of both the politicians and the whalers, as the scientists were left to determine catch limits according to a fixed procedure the Commission itself had endorsed (interview with Greg Donovan). Many imperfections with the procedure have since been brought to light, notably its flawed presumption of a near-perfect knowledge of initial stock sizes (Donovan 1995). Nonetheless, the NMP brought a new authority onto the SC such that, after 1974, governments were more hesitant to deviate from its recommendations. Furthermore, by devising a management system where stocks of whales could be managed on an individual basis, it defused the preservationist argument for a blanket moratorium (Gambell 1989). The SC had seized the middle ground and was slowly emerging as the new locus of power and authority and, consequently, as the bastion to conquer for those who wanted to see whales preserved rather than exploited.

The Revised Management Procedure: Science Multiplied (1975 et seq)

A New Global Context for Cetological Discourses

Outside the IWC, Stockholm had opened a watershed for environmental research (Jameson 1996). It had created a new context where two discourses conjoined: long-standing, slowly matured scientific discourses on problems of ecological degradation and the nascent activist discourse on "doing something about" the global environment. Because of its role in the rise of the global environmental discourse (see chapter 5), the whale signifier provided the nodal point for this articulation. In this context, cetological discourses began to proliferate beyond the confines of the IWC's SC. Thus the "Decade of Cetacean Research," which had been initially launched by the SC in 1975, was rapidly taken over by the nascent UN system of environmental knowledge production. In 1973, once IWC scientists had rejected the blanket moratorium, the FAO and UNEP together established an Advisory Committee on Marine Mammal Research (ACMMR) as a subsection of the FAO's Committee on Fisheries, for the purpose of undertaking an independent review of the stocks of marine mammals. Although no rigid walls were erected between the two institutions and the scientists readily went from one to the other (interview with Ray Gambell), the establishment of a parallel setting for whale research generated some measure of rivalry for recognition as the official, authoritative locus of production of cetological knowledge. And as they competed with each other in order to be heard, the voices grew increasingly louder. FAO–ACMMR and UNEP held a Scientific Consultation on Marine Mammals in Bergen in 1976 with Sidney Holt as convener. These consultation meetings spanned several years and yielded a four-volume edited collection, *Mammals in the Sea*, published jointly by FAO and UNEP between 1978 and 1981. A key purpose was to draft founding documents that would provide the basis for developing protective marine mammals policies around the world, modeled on the United States—the only country to have such policies—whose Marine Mammal Commission provided the templates, such as the nomenclature of species. One example of such a document is the IUCN/FAO/UNEP-endorsed marine species identification manual, *Marine Mammals of the World* (see Jefferson and Leatherwood 1993 for the latest edition), that was to provide the official classificatory scheme for future national marine mammal policies worldwide.[10]

More than any piece of science produced by the SC, the research compiled in these four volumes has provided the scientific backing for the preservationist stance against whaling. My argument is that this new discursive context—the conflation of scientific and activist discourses—had been necessary to being able to recast the research upon the animal and away from the requirements of its exploitation, which invariably framed any research undertaken by the SC, in accordance with its mandate. In other words, the question of whale intelligence was simply not one the SC could inquire into. It took this new institutional setting, in which alternative ways of knowing and talking about the whales could develop, to be able raise the question of whale intelligence and establish it as a legitimate object of scientific enquiry. One of the first times it appeared as a research topic was in a paper, presented at Bergen and included in the edited volume, by Jean-Paul Fortom-Gouin, a key anti-whaling activist with no formal cetological training who owned a dolphinarium in Florida, entitled "Some Aspects of Cetacean Neuroanatomy."[11] Nor did the SC's research remain isolated from these broader discursive shifts.[12]

The bold tone of the Bergen conference's Final Recommendations (also reprinted in the edited collection, 1978, 32), in which they address on an equal footing both "governments" and the "governing organs" of international organization (specifically the FAO, UNEP, and UNCLOS), reflected science's position of authority vis-à-vis policy makers with regard to environmental issues in this new discursive context. However, it also served to consolidate and extend it. Overall this research generated considerable policy outcomes. In 1978 the FAO and UNEP jointly launched the Global Action Plan for the Conservation, Management, and Utilisation of Marine Mammals, complete with a newsletter, *The Pilot*, and a popularizing pamphlet, *Marine Mammals*. Its aim was, in the words of the pamphlet (UNEP 1979, 37), to create a "framework for policy-planning and programs formulated by the international community," but it was also (UNEP 1979, 36) explicitly to "generate a consensus among the governments of the world on which to base a global policy for marine mammal conservation." Thus celebrating the emergence of a global scientific consensus on whales, these producers of cetological knowledge effectively constituted it.[13] The pamphlet (UNEP 1979, 37) continues, "For the first time world-wide concern for the state of these animals is being assembled into a single force, under the auspices of the United Nations."

This performative constitution of a global, informed consensus about whales carved out a considerable role for NGOs. For Bergen also served to establish the new part to be played by NGOs in the global structures of knowledge production centered on whales. NGOs such as the WWF, the Sierra Club, and FoE had all taken part at the conference, not merely as observers but as members of the "Scientific Consultation." Both the Final Recommendations and the Global Action Plan had hailed NGOs on par with governments as key actors in this area.[14] For knowledge to be effective it had to be channeled into popular awareness; NGOs were in a strategic position in this regard. Furthermore, the techniques to capture small whales and dolphins alive had been perfected in the late 1960s.[15] Against a backdrop of marine parks appearing across North America, the whale became an object of a new form of "pop science."[16]

A New Activist-Scientist Front in the IWC

Soon NGOs became involved in the production of cetological knowledge within the IWC. In 1977 the Commission voted to open the SC to "scientifically qualified observers" nominated by "any international organisation," including nongovernmental ones (IWC 1978, 27). This opened new channels between anti-whaling activists and the more "sympathetic" scientists. NGOs began employing scientists, either members of the SC or those closely associated with it, as consultants, commissioning work that could challenge official IWC science. At this point one man took on a critical role. Sidney Holt had been instrumental in drawing whale research out of the confines of the IWC and funneling onto it the momentum of Stockholm. Throughout the 1970s and 1980s he maintained his position as the critical eye of the SC, never quite in the front line of the science being done but always having a word of advice to dispense as an experienced and authoritative scientist. On the Committee of Four (during the 1960s) Holt had unambiguously endorsed the SC's founding mandate, conservation in view of exploitation, and worked to calculate sustainable yields. The emerging anti-whaling movement questioned the very purpose of the IWC—managing whaling—and therefore by implication the science tailored to these management objectives. Holt found himself in a strategic position to reorient IWC science toward the goal of preserving whales. Thus in the SC (not in the Commission), Holt opted for a backseat from which he could draw the strings; from there he became the center of a slowly forming scientist-cum-activist front.

In 1978 the Washington-based Whale Protection Fund contributed U.S. $25,000 to finance a reanalysis of the IWC's sperm whale computer models by Sidney Holt of the FAO and John Beddington of the IUCN (M'Gonigle 1980, 161). This was effectively the first piece of NGO science, and its results were presented at the SC meeting in 1978 to contest the sperm quotas devised under the NMP and adopted by the Commission the previous year. That year Beddington was instrumental in obtaining a zero quota for sei whales. Furthermore, a new agenda item was placed at the last minute on the SC's agenda (by Jean Paul Fortom-Gouin as the Commissioner for Panama; see next chapter), the "ethics of killing cetaceans," disrupting a meeting of the SC, which declared itself unprepared for the matter (M'Gonigle 1980, 165). In 1979 the People's Trust for Endangered Species commissioned John Beddington and his student Justin Cooke to carry out new research. The extensive work of William de la Mare and Holt, funded by the Environmental Investigation Agency (a Greenpeace offshoot), was instrumental in bringing to light the inadequacy of the NMP to the whole SC.[17] According to Michael M'Gonigle (1980, 179), who was then Greenpeace representative at the IWC, NGOs increasingly began channeling their funds into research that would undermine the NMP. Besides producing alternative research, that same year NGOs established themselves on par with the IWC by cosponsoring an IWC special meeting on Cetacean Behaviour and Intelligence held in Washington DC—then the heart of anti-whaling lobbying—which few SC members attended (IWC 1980).

The IWC Moratorium Resolution

The adoption of the ten-year moratorium on commercial whaling in 1982 entrenched a rift that had been slowly taking shape within the SC. A majority of scientists continued to oppose the blanket moratorium, advocating instead a stock-by-stock approach, whereas a dissenting group, including Chapman (who had been persuaded by Holt), de la Mare, and Holt, lodged a statement of support (IWC 1983, Annex M). More significantly still, this line of divide ran through the Committee of Four. Allen added his name to the moratorium critics, and Gulland began distancing himself from the IWC altogether. This division reflected a fundamental difference in the way the scientists conceived of their role. The majority of scientists continued to abide by their original mandate, which restricted them to determining sustainable catch levels, whereas a new group of scientists took on additional "ethical," anti-whaling concerns. Reflecting over the period 1980s–2000s, in a personal interview IWC scientist Doug Butterworth remarked to me how "old school," utilization-orientated cetologists were coming increasingly at odds with a new generation of scientists green to whaling (interview with Douglas Butterworth).

The moratorium resolution in 1982 was passed as a two-pronged initiative. First, the pause in commercial whaling was to allow a "comprehensive assessment" of whale stocks "by 1990 at the latest," leaving open the possibility of reestablishing catch limits, should whale stocks satisfactorily recover (IWC 1983). The spotlight was once again upon the SC, whose task was laid out for the next decade: to establish clear and reliable whale population estimates and to address the shortcomings of the NMP (Donovan 1989). The late 1980s saw the multiplication of "sighting surveys" to document the state of whale stocks. These research programs have been undertaken both collectively, under the auspices of the IWC's research programs—first the International Decade for Cetecean Research (IDCR) program, launched in 1978 and subsequently replaced in1996 by the ongoing Southern Ocean Whale and Ecosystem Research (SOWER)—and increasingly individually, by country members who then hand over results and methodologies to the SC. It is noteworthy that Japan, who is under increasing fire for its "scientific whaling," has in fact been a main contributor to IWC science over this period, both under SOWER, for which it has provided the vessels for the sighting cruises (IWC 1996), and under its highly contentious research program called Japan's Special Permit Research on Minke Whales in the Antarctic (known as JARPA). Japan's efforts were thus key to enabling the SC to establish its first estimates and thus carry out its mandate (interview with Ray Gambell). And to the surprise of anti-whaling activists, the first results in 1990 revealed that the Antarctic minke whale population of the Southern Hemisphere numbered 761,000 whales.

Since then SC's task has effectively developed into a long-term and ongoing undertaking, and so far only a small range of whale stocks have been assessed.[18] However, already with the first estimates, scientists were faced with hugely diverging results regarding the status of whale stocks, with some showing as highly endangered whereas others were proving healthy enough to sustain controlled catches. By way of comparison with the minke whales, estimates for the western North Pacific gray whale were established at 121 individuals in 2007 (up from under 100 in 2003). In the face of such diversity in whale populations, scientific efforts were directed toward developing a fine-tuned, stock-by-stock approach. By contrast, as the scientists had already expressed when the NMP was adopted, a management approach that imposed a sweeping moratorium on all species was simply at odds with such diversity. What the anti-whaling consensus had said was that there was a linear decline in whale stocks; what the science showed was that there was not; and yet the consensus did not change as a result of the science's improving.

The second part of the SC's original 1983 mandate, exploring other conceptual approaches to management, was addressed through what developed into a form of open competition. In 1989, five different groups of scientists put forward five proposals for alternative management schemes: two Japanese, two Icelanders, two South Africans, and two scientist–activists, Justin Cooke and William de la Mare, who each presented a proposal. In a climate of increasing polarization in the IWC, the SC established a Management Procedures Sub-Committee for the purpose of weighing each proposal on its scientific merits and, indeed, of ridding it as much as possible of "unscientific" motives. Each one was tried and tested by a common preagreed computer simulation process. Eventually two proposals were selected on their scientific merits: the South-Africans Punt-Butterworth's and Cooke's. Both these proposals were judged to perform similarly well; their main difference was in their design. One placed slightly more emphasis on catches, according to the traditional IWC mandate, and the other emphasized minimizing risks. Thus the final decision would be a choice between these two criteria. In the end, at a time when the precautionary approach was becoming entrenched into international law and decision making (Sands 1995, Hajer 1995), the SC opted for the more precautionary approach, Justin Cooke's.[19] The final SC report recommended the Cooke proposal as the "best" management procedure to the Commission, and its catch limit algorithm (CLA) became the core of the RMP (once rules to handle multiple-stock situations were later added).

Adopting the Revised Management Procedure

In an unexpected turnabout, Justin Cooke at the last minute joined a minority group, again revolving around Holt and de la Mare, withdrawing his support for his own work and backing instead de la Mare's proposal (IWC 1992, 56; Schweder 2000). This proposal, while emphasizing the precautionary approach (risk rather than catch), if passed, would have terminated all minke whaling in the North Atlantic, while allowing some catches in the Southern Ocean (because of some inbuilt "multi-stock"/ocean basin rules). In other words, politically, it would have divided the whaling camp, between Iceland and Norway on one hand, and Japan on the other, who was promised some Antarctic minke whaling. The Southern Ocean Sanctuary proposal that was passed two years later (see the next chapter) would have blocked the latter possibility. However, at the time, the anti-whaling camp could not bet on the success of this move. In addition, Norway was then the most pressing target for anti-whalers, for it had already adopted the RMP and stood ready to resume whaling; it was only waiting for the new procedure and results to be laid out to the Commission. Cooke's proposal, by outperforming de la Mare's, had effectively ruined their plan. Whether any peer pressure was exerted from fellow activist–scientists is difficult to assess. Whatever Cooke's motives, they must have been strong enough for him to vote against his own work. One scientist who attended most SC meetings throughout the period summed it up by explaining that, until the RMP had been finalized, given the numerous uncertainties, the anti-whaling scientists had in fact been putting forward "reasonable cases" in objecting to quota calculations and thus constructively fueled the scientific debate. Once the RMP was completed, however, it became clear that they had shifted to obfuscation tactics. Moreover, they surfaced again in another incident in 1996, in the discussion of the Northeast Atlantic minke whale stock estimate derived from a Norwegian sighting survey. Cooke agreed to this estimate at an SC intersessional meeting. However, this estimate, which established the stocks as healthy (118,000 whales), effectively condoned Norwegian whaling. Right at the close of the 1996 SC plenary, which had more public "visibility" than the intersessional subgroup meeting, Cooke "reneged" on his agreement. This threw a spanner in the works of a tight meeting program, whose recommendations needed to be passed on to the Commission, and Cooke came under fierce criticism from the other scientists for his obstruction (IWC 1997, 73–74 and 76).

When the completed RMP was put forward to the Commission at the 1993 annual meeting on time for the review of the ten-year moratorium on commercial whaling, it was rejected outright by the Commission, who by then counted a majority of anti-whaling states (see chapter 7). Iceland resigned in protest. Less predictably, the Chairman of the SC, Philip Hammond, who had been painstakingly trying to maintain a middle ground in an increasingly polarized SC, submitted his resignation to the IWC with the following words:

> What is the point of having a Scientific Committee if its unanimous recommendations are treated with such contempt? . . . The mechanism for safe whaling . . . a unique piece of work for which the Commission had been waiting for many years [had been left hanging in the air]. The reasons for this were nothing to do with science (quoted in Blichfeldt 1994, 4).

The Commission subsequently adopted the recommended RMP the following year (IWC 1994, 43–44). At that point, a distinction was drawn upon the Commission floor between the RMP and a Revised Management *Scheme* (RMS), which was to solve all the other, nonscientific management issues (such as quota enforcement and codes of conduct for vessel surveys; IWC 1994, 43–44). In a sense this distinction has helped to shelter IWC science from the increasingly polarized IWC debates by carving a separate place where all the contentious, political issues could be relegated. However, it has also meant that, for all these efforts deployed in the early 1990s, IWC science has since remained underused and increasingly disinvested, since the RMP has been shelved until completion of the RMS. The problem is that, since 1994, the RMS has turned into an ever-expandable list of issues that are unlikely to be resolved in the current context, and its completion is thereby perpetually postponed. For example, the United Kingdom has added animal welfare in 2000, and New Zealand has called for international supervision of domestic markets under a DNA monitoring program, a proposition to which Norway, which already operates its own DNA monitoring system, is unlikely to yield. In fact, perpetually maintaining this list open has become a key strategy for upholding a *status quo* that effectively benefits *both* the anti-whalers and the whalers (Ishii and Okubo 2007). It is of course essentially an anti-whaling status quo, since the commercial whaling is perpetually suspended. However, insofar as whaling is actually increasing on the ground, the whalers have not been significantly hindered either, and they have, moreover, developed new discursive strategies around it (see chapter 10). Meanwhile, beyond the IWC, the RMP has been put to use to regulate commercial whaling at the domestic level in Norway (since 1993) and briefly in Iceland;[20] but also in the IWC's own aboriginal whaling schemes (see chapter 7). It has also inspired the development of management procedures in fisheries around the world (personal correspondence with Philip Hammond).

Conclusion

Returning to the question with which this chapter opened, the whaling case has shown that when political differences run deep, science does not have the power to provide a rational basis for the development of a common understanding, and from there, of successful collective policies. Despite improvements in IWC science, IWC debates became, overall, increasingly polarized. This is because scientific discourses form part of the broader discursive order. Insofar as scientific discourses do not develop in isolation from other discourses, this discursive order sets constraints upon science's autonomy, what questions it can ask, and what research it can undertake. As a result, science does not provide the means to step out of that particular discursive order so as to develop a consensus that can resolve political differences, when these differences are in fact constitutive of that order—that is, when they reflect the dynamics of a dominant discourse. What science does not have, above all, is the power to *make* people and states *listen*.

Hence science cannot overwrite a fully entrenched dominant discourse. When its conclusions run counter to the expectations inscribed in a dominant discourse, they are simply overlooked. Thus in the pre-1964 whaling order, despite the creation of an international regime that conferred a key role upon the science, scientists could not tell whaling policy makers what they did not want to hear—that some whales were rapidly disappearing. Conversely in the anti-whaling order, scientists could no longer tell (a majority of) anti-whaling policy makers what *they* did not want to hear, namely, that certain stocks of whales might not be so endangered. For the implication is that they could sustain some measure of controlled exploitation, a possibility that is simply precluded by the anti-whaling discourse. In fact, the anti-whaling discourse does more than foreground the scientific question of whales' endangeredness. It answers that question in the affirmative. A key articulatory practice of the anti-whaling discourse is not only to associate any evocation of the whales with the powerful notion of their endangeredness, as we saw in the previous chapter, but it is to provide the automatic answer to the underlying question, such that they are assumed to be endangered prior to the question having actually been asked. Thus a key articulatory effect of that discourse is that it forecloses the scientific question of their endangeredness from the onset.

This does not mean however, that science is entirely reducible to the dominant discourse, as a mere—albeit slightly more rational—by-product. Not only would this be denying the specificity of scientific discourses but it would amount to denying scientists their agency and the possibility for science to develop from the inside. It would also be committing the Foucauldian fallacy of evacuating agency altogether. The whaling case has shown that, although science may have started off with very little autonomy in the whaling order, scientists were subsequently able to institute some measure of autonomy, vis-à-vis policy makers, but also vis-à-vis anti-whaling activists, who had by then become key players in the whaling regime, as we shall see in the following chapter. They achieved this by taking advantage of the instability wrought by the advent of a new discursive order. The period 1973–1993 thus saw the rapid development of whale management science, which provided a model for fisheries management at large.

What specific powers of persuasion does science have, then? Why, despite the fact that anti-whaling activists increasingly turned away from the official IWC science to develop their own, did they continue to rely so centrally on science? Why, in other words, did the resort to science continue to remain a key discursive strategy? At this point, it is necessary to distinguish between IWC science and the broader power of science in modern societies—for activists shunned the former and tapped into the latter. Modern scientific discourses inherently hold a particularly powerful subject-position. This subject-position, which is a position of authority, is what anti-whaling activists continue to vie for. Science is powerful because it constitutes a discourse of authority in modern societies. As Stanley Aronowitz (1988, vii–xii) has shown, this power, in turn, relies on a particular conflation of "knowledge" and "truth." In other words, science has thus successfully laid claim to the monopoly over the production of truth. From there stem claims to detaining the only legitimate knowledge about the world, which, in turn, form the basis of the claims to authority in the modern world (Aronowitz 1989, ix). Yet this particular way of knowing may prove to be as time bound and situation specific as historically prior knowledge/power complexes.

Anti-whaling activists thus sought to occupy a subject-position laden with the power of authority and legitimacy. They did this by starting to produce their own whale science. NGOs began multiplying science on the margins of the IWC as a way of challenging this centralized, authoritative science of the IWC. They thus successfully initiated a sort of decentering of subject-positions, a dispersion of the possible places from which (new types of) knowledge about whales could be proclaimed. They effectively democratized the production of knowledge by appropriating it on the margins of the whaling regime. These anti-whaling activists claimed *authorship* of the scientific discourse on whales. That is, they brought this power of science onto their subject-position by successfully establishing themselves as authors not only of an activist (partisan) discourse but of the *authoritative* discourse on whales itself, that is, of the scientific discourse. Hence they established themselves in the position to *authorize* what could be known about whales, in addition to what could be said and done with whales.

However, this proliferation of IWC science also led to its fragmentation. The whaling case has shown that there is not one disinterested and autonomously developed science driving policy making but many different forms of knowledge about whales, or local cetologies, which sometimes contradict one another. Nowhere is this better brought home than in the experience of managing whaling on the ground. For instance, in the 1970s, a U.S. scientific survey was carried out to determine a population estimate of Alaskan bowhead whales from which the Inupiat aboriginal catch quota would be derived. In 1977 the scientists concluded from their findings that there were fewer than six hundred whales left and the population was further declining, which resulted in an IWC ban the following year (see chapter 7). Yet the Inupiat whalers rejected the figure, contending that the whales were hiding under the ice. Subsequent surveys and analyses vindicated their argument, which had been based on experience and intuition rather than algorithmic estimates, establishing the number of whales was at least double that proclaimed by the scientists, with an average population increase of 3.1 percent per annum (Hess 1999, 12). Indeed, these different ways of knowing whales are social and political constructs, which sometimes drive policy-outcomes regarding whales far more than the science can support.

7

The Anti-Whaling Campaign

Whereas chapter 5 analyzed the broader context in which a new discourse on whales took hold, this one closes in on the anti-whaling campaign itself. From being normal and unquestioned in the early 1960s, in less than two decades commercial whaling became unacceptable and outlawed by an international moratorium. This chapter examines in detail with what precise stunts, gestures, tactics, and strategies activists successfully denormalized whaling. Hence the analysis is pitted at one level lower, at the points of contact where change was ushered in. Its time span is the ten years leading up to the 1982 IWC moratorium on commercial whaling. What activists achieved with this new type of campaign deployed at the supranational level was the first global, precautionary suspension of a natural resource's commercial exploitation. This episode thus constitutes a milestone in the history of activism at large, as activists honed here many of the skills and strategies that have come to constitute the arsenal of contemporary global activism, beyond environmental issues. With this campaign activists effectively created a new type of global political space, complete with its own set of rules. The movement of the chapter follows that of the campaign itself: In the first part, the stage was the globe as a whole, as the aim was to win over the heats and minds of the entire world. The second part focuses on the conquest of the inner bastion, the IWC.

The Globe as a Battleground

The Actors

The anti-whaling campaign was not launched as a carefully planned enterprise; rather, it evolved out of disparate moves, one-off actions, and publicity stunts by a handful of bold activists that were only progressively coordinated into a campaign. Individual actors played a key role in reframing whaling. Their strengths were their media savvy, an acute flair for political opportunities, and an ability to transform these openings into a long-term strategy. I begin by portraying some of the main "whale warriors," to use activist David Day's (1992) term.

Joan McIntyre, who first coined the anti-whaling synecdoche examined in chapter 5, exemplifies the hands-on activism that was key to drawing the world's attention to the plight of the whale. She left FoE to found the first NGO dedicated solely to saving whales, Project Jonah. She saw the opportunity at the Stockholm conference, of which she recounted: "I realized that the press was the greatest resource there and they had nothing to write about. So I organized a big outdoor rally and a whale walk" (*LA Times* 1974).

She convinced Maurice Strong to speak at the rally and proclaim the whale the symbol of the UN conference. From there, she launched a worldwide campaign coordinated from Project Jonah's base in San Francisco—formerly the main whaling port in the North Pacific. It had soon stretched an arm into Europe, Australia, and New Zealand. Rapidly eclipsed by other organizations latching onto the cause, it was nonetheless one of the most influential organizations at the takeoff of the anti-whaling movement. It shaped the single-issue activism that still characterizes anti-whaling NGOs today. The success of the campaign intensified personal rivalries between the different hands-on groups over who owned the issue and, notably, between Project Jonah and Greenpeace (Hunter 1980, 233).

Christine Stevens represents a different type of activist altogether. Wife of an industrial magnate-cum-Broadway-producer, well connected in Washington, notably to the Kennedy family, Stevens was known as the mother of the animal protection movement. She founded the Animal Welfare Institute in 1951 and its Washington lobbying arm, the Society for Animal Protective Legislation, in 1955. Mrs. Stevens held a salon where any young politician aspiring to the White House was to be seen. She is personally accredited with the passage in Congress of "more than a dozen landmark animal protection laws," notably the 1966 Animal Welfare Act (Saxon 2002). For some it was not McIntyre but Stevens who launched the whaling camp in 1971. She provided the funds for Craig Van Note of the Monitor Consortium, an association of NGOs, to publish the anti-whaling NGO newsletter *Eco*.

Within the IWC, a steering committee orchestrated the moves that led to the moratorium. Its key members were David McTaggart, the charismatic Greenpeace leader, Sir Peter Scott of the WWF, who was personally dedicated to the issue and became advisor to the United Kingdom government in 1979, and Cornelia Durrant of FoE. A common activist practice, developed in the late 1970s and still applied today, was to use IWC meetings to cultivate close ties with chosen individuals on country delegations. Activists also began donning several caps at once, that of financiers, scientific advisors, even country Commissioners. One such actor was "Doctor Lyall Watson, para-psychologist, guru and author" (Day 1992, 21), who had written a book on whales, *The Whales of the World* (1981), and played a key role on the Seychelles delegation. He brought the funds of the Threshold Foundation to the cause, a charity founded by the brother of the Iranian Shah Pahlavi for the purpose of fostering "world peace" through "mutual understanding." Another key figure was Jean Paul Fortom-Gouin, a Bahamas-based Frenchman who owned a dolphinarium in Florida, had his own NGO, the Whale and Dolphin Coalition, and authored scientific papers on cetacean intelligence for the FAO Bergen conference as we saw in the previous chapter.

In addition to activists-turned-scientists, some of the most important moves were initiated on the scientific front by scientists-turned-activists. Dr. Sidney Holt, after retiring from the FAO, became one of the strongest pillars of the campaign, not least because of his intimate knowledge of IWC workings. This tandem of McTaggart and Holt soon became the main axis of the campaign. Other activist–scientists included de la Mare, member of the IUCN, who worked with Greenpeace International, and Justin Cooke, also affiliated with IUCN, WWF, and employed as consultant for various NGOs. Later came Michael Tillman, member of the SC and Alternate-Commissioner for the United States, who was rebuked by the Animal Welfare Institute for initially working toward the RMP in 1992 and then eventually awarded the Institute's Schweitzer medal for "preventing the adoption of the RMS" the following year (High North Alliance [HNA] 1994b). Beyond the IWC other high-profile scientists brought their name to the cause, such as the biologist Roger Payne, famous for bringing the *Songs of the Humpback Whale* to the public.

Spheres of Activism The anti-whaling campaign carved out several spheres of activism. These actors formed a core of activism powering the campaign, right through to the suspension of commercial whaling. Around this inner circle formed another, looser circle of individuals drawn in at particular points in the campaign as well as groups created to tackle specific aspects. The main target for these self-proclaimed disciples of Marshall McLuhan (Pearce 1991, 19) who were often media professionals themselves (notably the Greenpeace founders; Hunter 1980), was a growing global media audience—the "global village." Thus beyond first two spheres of "active" activism, the anti-whaling campaign modeled new varieties of a more *passive* activism. This was driven by two ideas: first, turning everyone into an "eyewitness" of the whale's plight, drawn from the Quaker tradition that the Greenpeace activists specifically had brought to the campaign (see chapter 5), and second, breaking down the human/animal barrier that was seen as the main source of human indifference to whales (D'Amato and Chopra 1991). Early on NGOs began multiplying opportunities for the public at large to engage with the plight of the whale. These included, on the more participatory side, whale marches, as well as letter-writing campaigns that performatively generated the "voice of the future," mobilizing children as key symbols of universality (Einarsson 1993). Important too was the emphasis on new ways of interacting with the whales, in activities such as whale-watching. In this way, simply paying one's membership fees to an NGO, or even merely watching the whale's plight on the evening news, became a way of taking part in the anti-whaling campaign. An important achievement of the anti-whaling campaign was therefore that it cast the global consumer of images as an anti-whaling activist (see chapter 9). This was key to establishing a dominant global discourse on the ground.

The Actions

Carving Out a New Space for Activism Anti-whaling activists effectively carved out a new type of political space, beyond the traditional space of the state: what Thom Kuehls (1996) calls the space of ecopolitics. Kuehls analyzes the practices of sovereignty as a process of constantly reasserting control over a territory—an ongoing *reterritorialization*. Environmental activism, by contrast, developed an alternative, *deterritorializing* politics such that "the extension of [the movement's] support increases its territory as it deterritorializes the state" (Kuehls 1996, 50). By staging their confrontation upon the high seas—traditionally the whalers' space—anti-whalers had displaced the locus of political struggle onto that smooth, asovereign space beyond the nation-state and its associated particularistic nationalisms. At the same time, this traditionally apolitical space was "reterritorialized," in that it was reappropriated, invested with new political significance. Activists designated the globe as the new space for political mobilization, for which the whale, whose migratory routes span the entire globe, provided here again an adequate symbol (see also Epstein 2003, Yearly 1996).

Riding the airwaves of a globalizing media was key to carving out this new political space. As we saw in chapter 5, shock tactics and visibility were central to their strategy to denormalize whaling, which former Greenpeace United Kingdom program director Chris Rose (1993, 287) captures as "more an imagology than an ideology." Indeed, many of the anti-whaling activists had a keen understanding of the power of images. Having themselves become "electronic age celebrities" overnight (Hunter 1980, 232), they rapidly knew to wield fame to cultivate long-term relationships with the media and thus maintain the world's attention upon the whale. Involving the celebrity of the day has since remained a regular NGO practice: from Gregory Peck, the actor who had played Captain Ahab in the 1950s, to Leonardo di Caprio and even James Bond hero Pierce Brosnan.

The second component of this long-term relationship with the media corresponds to what Keck and Sikkink (1998) have identified as "information politics." By systematically providing otherwise unavailable information, NGOs successfully positioned themselves vis-à-vis the media as alternative and authoritative sources of information. In this case, however, activists wielded their information politics not just to change the hearts and minds of the global public at large (and thus to pressure states indirectly) but also to act directly upon the whaling regime. The unreported catches from "pirate whalers," whose vessels eluded the IWC Inspection and Observer Scheme, severely undermined the ability to determine catch quotas. Activists-turned-ecodetectives launched vast fact-finding missions to uncover the unreported catches. They then exposed their findings on television, notably through the airing on the British Thames Television of a documentary on the operations of the *Sierra* in 1979 (Day 1992). Whereas in the days of Onassis's *Olympic Challenger* the struggle to obtain compliance was waged in the secluded rooms of (generally Norwegian) courts, it was now brought in full view of the public. The evidence compiled by activists was simultaneously brought to the IWC's Infractions Sub-Committee. Similar ecodetective missions were repeated, notably against the USSR. Through their dramatic use of information, gathered on the sea and diffused on the airwaves, NGOs effectively carved out for themselves a new and powerful monitoring function on the margins of the official whaling regime.

Displacing the Legal/Illegal Boundary From this new space they had carved out activists were able to displace the boundaries of legality/illegality and, correspondingly, of the acceptable/unacceptable. This displacement of legal and discursive boundaries was thus central to denormalizing whaling. The trajectory of the term "illegal whaling" through the whaling regime is illustrative in this regard. "Illegal whaling" first surfaced in activist discourse to denounce the whaling of non-IWC member countries. Only "pirate whaling" was, strictly speaking, illegal, since the vessels were often traceable to IWC member countries (the *Sierra*, for example, was tracked down to South Africa). Non-IWC members were not bound by IWC whaling regulations, and their whaling stood beyond its remit, technically "alegal" rather than "illegal." This unbounded whaling, which amounted to 13 percent of world catches in 1978 (M'Gonigle 1980, 191), remained an important impediment to controlling whaling. Activists began deploying similar missions against whaling operations in nonmember countries—for example, in Taiwan in 1979, in Peru in 1978–79, and in Chile in 1981 (Day 1992, 78–81). These expeditions, also relayed by the media, cast this form of whaling on par with pirate whaling, and effectively, by association, as illegal. Interestingly, the shaping of this new discursive category of "illegal whaling," which rapidly took hold in the international media, reinforced the legal regime, for it brought additional nonmember states to the IWC table. Chile, Peru, and Spain joined in 1979, also as a result of pressures from the U.S. government.

A more far-ranging displacement of the boundaries of legality/illegality was operated by "ecoterrorist" practices inaugurated by Paul Watson, a former Greenpeace activist who had split with the group over its "moderate" methods. Dissatisfied with the outcome of the *Sierra* scandal, he decided to put an end to its pirate whaling himself. In July 1979, outside the port of Leixos, Watson rammed the concrete-reinforced steel bow of his *Sea Shepherd* into the *Sierra* to tear off its harpoon, swung round on itself, and tore into the vessel mid-ship, in full view of Portuguese port authorities. Watson was arrested by a Portuguese destroyer. After court hearings, however, his ship was held in lieu of damages and cost claims on behalf of the *Sierra*, but Watson himself was surprised to find that no criminal proceedings were taken against him (Day 1992, 57). He left Portugal with impunity.

These extreme direct methods had been used against a dubious whaling boat; in a sense both belonged to the realm of illegality. However his next stunts placed him squarely in opposition to national law. When Norway announced its decision to resume whaling in 1992, he turned to Norwegian whaling vessels. On Boxing Day 1992, in the Lofoten Islands, Watson boarded the empty *Ny-braenna* and opened the taps in the engine room to flood the whaling vessel. The Norwegian courts sentenced Watson in absentia to 120 days in prison. Watson was arrested in the Netherlands on his way back to the United States on an Interpol warrant in April 1997 and held for custody for 80 days. His arrest caused a popular uproar amidst animal welfare activists, who launched an unrelenting campaign for his liberation. The Dutch judges rejected Norway's extradition request. Watson was only fined for falsely emitting distress signals. This time the vessels Watson had destroyed were lawful and in fact tightly controlled by their own government (see chapter 10). A criminal according to one country's legislation, Watson never served his sentence, and his widespread popularity remained untainted. In April 2003 he was elected onto the board of directors of the Sierra Club. The anti-whaling campaign not only rewrote whaling as "illegal," but it succeeded in establishing its more criminal deeds as "acceptable."

Investing National Laws: The United States National spaces were also a key target for activists. The United States was a natural starting point, not only as the birthplace of the anti-whaling movement but because the lobbying culture and characteristic openness of its political debate made it especially accessible to activists. An added benefit, perhaps more intuited than known, was the sort of structural advantage that might accrue to actors belonging to the international system's hegemon (Josselin and Wallace 2001). Anti-whaling activists became cunning lobbyists and positioned themselves so as to weigh in at key points in the political process (see also DeSombre 2001). For example, Christine Stevens took part in the congressional debate leading to the 1972 MMPA, resting her case with Roger Payne's *Songs of the Humpback Whale*.[1] The various whaling moratorium resolutions presented by the U.S. delegation throughout the 1970s at the international level had been either initiated or carried though Congress by various NGOs—the first one in 1972, for which the Animal Protective Legislation and the International Society for the Protection of Animals had played a key part.

Besides endangered species laws, another line of legislation, namely (and somewhat ironically), fisheries law, was invested with whale protective capabilities. Two measures were key. First, in 1971 the Congress enacted the Pelly Amendment to the 1967 Fishermen's Protective Act. This conferred upon the president discretionary authority to ban imports of all fishery products from a foreign country when he determined that the nationals of that country were "conducting fishing operations in a manner or under circumstances which diminish the effectiveness of an international fishery conservation program," which explicitly extended to cetaceans (U.S.C. 1975). The president could invoke trade sanctions upon certification by the secretary of commerce that a conservation program was being hindered, even if no species or stock was in danger of extinction and no treaty violated. Although the stated goal was multilateral, the actions considered to diminish the effectiveness of such programs were determined unilaterally (Charnovitz 1995). The second law, the 1979 Packwood–Magnuson Amendment to the 1976 Fisheries Conservation and Management Act, was the brainchild of the NGO Defenders of Wildlife (M'Gonigle 1980, 201). Indeed, Senators Packwood and Magnuson both had long-standing ties with conservation groups, to whom they paid explicit tribute in presenting the law. It required that nations found to be acting against an international fisheries conservation agreement be excluded from fishing within the 200-mile U.S. Exclusive Economic Zone (EEZ). Domestic legislation was thus mobilized to reinforce and extend the international regulatory regime.

An interesting point to note in the application of these laws is the complementarity between state and nonstate measures. The 1974 NGOs' boycott of Japanese and Soviet goods (see chapter 8) was timed to coincide with President Ford's threat of a Pelly trade embargo on Japan and the USSR for objecting to a minke whale quota. Although the embargo was never implemented, the looming threat impacted negotiations at the IWC that year. Furthermore, NGOs would weigh into the process by pressing the U.S. delegation at the IWC to accelerate the process by which a whaling country was reported back to the president for sanctions. Moreover, as the Pelly Amendment did not require that the "offender" be party to the conservation program, it could be used to pressure non-IWC countries to join the whaling regime.

How effective were these unilateral trade measures in extending whale protection around the world? With U.S. $225 million worth of fish catches in the U.S. EEZ at stake, the threats held sway over Japan (M'Gonigle 1980, 190). However, while Japan had an obvious material interest in aligning its behavior, given the resources at stake, these interests only partially account for the way in which other countries reacted to U.S. measures, because countries who were not under similar material constraints behaved the same way. In other words, they adopted similar patterns of behavior despite no such material constraints. Both Korea and Peru joined the IWC in 1978 and 1979, indeed after being similarly threatened with sanctions (Scarff 1977). However, many more countries joined the IWC without having been directly threatened, such as Chile and Spain in 1979 or China in 1980.

Moreover, these laws proved more effective as a threat, that is, an effect of discourse—a speech act—than as actual sanctions (Austin 1962, Searle 1969). For few sanctions were actually implemented. It is useful here to distinguish between certification (when a country was established by the secretary of state to have violated conservation obligations) and sanctions (when the president actually implemented sanctions against that country). Certification was thus the point at which NGOs could exert impact. While the obligation to report the country was mandatory, the president's decision to sanction was discretionary. In actual practice, no state was ever sanctioned under Pelly for failing to accept whaling regulations. As for Packwood–Magnuson, the sanctions were applied twice only, against the USSR (whose commercial whaling was already being phased out) in 1985 and against Japan for its scientific whaling program in 1988. Nonetheless, in the first two decades, these U.S. legal instruments substantially impacted whaling practices at large.[2]

With a country like Norway, however, who had no fishing allocations in U.S. waters, Packwood–Magnuson sanctions were moot. When it unilaterally resumed whaling in 1993 (under its reservation to the moratorium), it was certified, but, much to the dismay of NGOs, President Clinton opted for "a continuing dialogue" instead of sanctions. Moreover, the discursive effectiveness of the threats diminished over time. Their credibility was undermined by the inconstancies and double standards that were taking shape as revealed by a rapid overview of the different cases. While Japan was actually sanctioned for its scientific whaling, Norway was never even certified, yet it never suspended its scientific whaling over the same period. Spain, for example, was not certified for objecting to quotas in 1980. Likewise, Canada, who quit the IWC in 1981, one of the more important whaling countries, was eventually certified in 1996 for whaling outside the IWC but never sanctioned. By roughly 1988, the threats lost their clout altogether (DeSombre 2000). Since then Japan has been further twice certified under Pelly for its scientific whaling (in 1995 and in 2001), to no effect. Iceland was certified for resuming its scientific whaling in 2003, equally to no avail.

The Victory Down Under: Australia and New Zealand Australia and New Zealand, also countries with relatively open political structures, were the next two targets to be roped into this English-speaking anti-whaling alliance. Both countries' waters were abundant in whales, as they are situated on whales' migratory routes to Antarctic breeding and feeding grounds. New Zealand waters in particular teemed with whales, due to a combination of oceanic currents, sand banks, and shallow coastlines (Gillespie 2001). With this unique set of geological features, its beaches are commonly the site of whale strandings. Although no evidence of full-fledged whaling traditions has been uncovered, in these regions whales were likely to be used on an opportunist basis by indigenous societies traditionally turned to the sea. In New Zealand, the Maori were known harvesters of marine mammals and "passive harvesters" of beached whales (Coutts 1976; Te Ohu Kai Moana 1999, 2). For whale strandings were seen as a *tohu pai* (a good sign), a gift from *Tangaroa* (the God of the sea), from which the whales themselves descended (Te Ohu Kai Moana 1998, 3). In the absence of land mammals, whales constituted a bounty: the meat was traditionally taken for food, oil was taken for preserving food and wood, and teeth and bone were intricately carved into ornaments and weaponry. The Maori became rapidly involved in the whaling operations developed by the Europeans (Morton 1982). These fertile waters were regularly visited throughout the seventeenth and eighteenth centuries by French, Dutch, Portuguese, Danish, British, and Yankee whalers. The competition for whales was an important motive in securing British sovereignty over the Islands (Belich 1996). Furthermore, whales and whaling grounds was one of the issues in the settlement process between the Maori and the British Crown entrenched in the 1840 Treaty of Waitangi (see chapter 11). Finally, Australia's and New Zealand's location was advantageous for whaling in Antarctica, the richest whaling grounds of the globe. An important whaling industry rapidly developed in both countries after the Second World War. By

1949 Australia harbored vast whaling plans. New Zealand whaling peaked in 1959–1960 when a total of 681 whales were caught (Tonnessen and Johnsen 1982, 547–549). Both concentrated on land-based whaling, essentially of humpbacks.

New Zealand's stance in the IWC had traditionally been close to Norway's in that both countries were the strongest advocates of a tightly restricted and controlled whaling. Like Norway in 1959, it has resigned when other whaling countries proved unable to accept the need for national quotas. However, when it rejoined the IWC in 1976, it was to bring commercial whaling to an end once and for all. New Zealand had been one of the countries where the NGOs' movement took hold most successfully, through Project Jonah but also through Greenpeace and FoE to a lesser degree. By 1978 New Zealand had adopted the MMPA (see chapter 11); foreign minister Talboys explicitly credited NGOs with the government's turnabout (M'Gonigle 1980, 188).

Australia would not so easily relinquish its whaling stance. Indeed, the 1974 NMP proposal had been put forward by Australia as a whaling country concerned with its good management. The delegates to the 1977 IWC meeting hosted in Canberra had thus flown to a whaling country. Yet as the last English-language whaling nation, Australia was an important bastion to conquer for the anti-whaling movement. Already for some years Project Jonah had been unrolling its dual strategy of targeting the public and lobbying government. In the run-up to the 1977 conference Fortom-Gouin's Whale and Dolphin Coalition raced the vessels of the Cheynes Beach Whaling Company with their dinghies off the Western Australian coast. A harpoon was fired, hit the propeller, and flipped the dinghy over into shark-infested waters; this dramatic confrontation was broadcast world wide. These stunts were matched on land by daily demonstrations throughout the conference. A 40 foot long white inflatable whale was pumped up into the skies opposite the conference venue. A mascot of many an IWC conference to come, Willie the Whale was born. The whaling conference had been successfully used to draw the world's attention to Australia's whaling practices.

In less than six months Australia would join the ranks of the whaling-redeemers-turned-fervent-advocates of the moratorium. An important figure in Prime Minister Malcolm Fraser's 1977 election campaign was his eleven-year-old daughter Phoebe, who wore a save-the-whale badge. He confided to Project Jonah lobbyists that he was "coming under pressure from home to stop the killing of whales" (Day 1992, 19). After the

1977 IWC conference an opinion poll had revealed 70 percent of Australians had been won over to the anti-whaling cause. He commissioned an independent parliamentary enquiry to look into Australia's whaling policies, where anti-whaling activists testified extensively. Local NGO pressure was relayed abroad with, for example, FoE instituting a boycott in the United Kingdom against companies using Cheynes Beach sperm oil (M'Gonigle 1980, 166). In November 1978 this last Australian whaling company announced its operations were closing due to financial difficulties. Whaling was outlawed by April 1979. In a dramatic historical turnabout, Australia became the first nation to shift the grounds of its policies from scientific considerations of levels of endangerment to ethical arguments about the intrinsic value of the whale. Fraser himself had declared, "The harpooning of these animals is offensive to many people who regard killing these special and intelligent animals as inconsistent with the ideals of mankind" (Day 1992, 19). From there, Australia projected its stance at the global level. Fraser continued: "Australia should pursue a policy of opposition to whaling and this policy should be pursued both domestically and internationally through IWC and other organisations."

By 1980 Australia had adopted the Whale Protection Act, the first piece of legislation tailored exclusively for whales. Just over a decade earlier, we were in a world where whales were tradeable commodities. We were then in a world where whales had become endangered; now, we were moving into a world where they had become ethical imperatives (D'Amato and Chopra 1991; see also chapter 10).

The Inner Circle: The IWC

By the early 1970s, activists appeared to be winning the battle of words. Yet while the discourses about whales had begun to shift around the world, the legal regime had not. Once the first whaling moratorium foundered at the IWC in 1972, it became clear that, to lock in the anti-whaling order completely—to close the gap between discursive and legal orders—the ultimate stronghold to be taken was the IWC itself. This section details the series of activist moves deployed within the IWC. There activists started gaining ground after 1978. In the first part of the decade, this advance was stalled by tensions within their most important ally, the United States, who had to accommodate a symbolically important whaling constituency. These two periods will be reviewed successively. First, I examine how the contradictions in the positioning of the United States were eventually resolved through the creation of a discursive distinction, "aboriginal versus commercial whaling," and by entrenching this distinction into international whaling law. I then overview the intensification of state–NGO interactions after 1978, building up to the ultimate anti-whaling victory, the proclamation of a moratorium on commercial whaling in Brighton in 1982.

The Moratorium on the Table The U.S. proposal for a whaling moratorium, which had been adopted unanimously by a 52–0 vote in Stockholm, received only four votes out of fourteen votes at the IWC (the United States, United Kingdom, Mexico, and Argentina). However, the following year, when Japan and the USSR rejected the year's whaling quotas, President Ford threatened a trade embargo under Pelly. For the first time, flouting IWC quotas had become costly for whaling nations, signaling the beginnings of a change in the dynamics of cooperation. When the moratorium was put forward anew at the 1973 vote, the four yes votes were joined by three former abstainees, France, Australia, and Canada (the latter two still whaling countries), and one former no vote, Panama. However, since the proposal qualified as an amendment to the Convention Schedule rather than a resolution, it required a three-quarter instead of a simple majority. The next nine years would thus be a relentless struggle to obtain this three-quarter majority.

At the 1974 annual meeting the moratorium issue was introduced for the third time. This time it was eclipsed by the Australian proposal ushering in the NMP. The NMP had struck a balance between diverging demands. Once a procedure was in place for systematically determining which populations were most in need of protection, a majority of countries chose to accept the recommendations of the SC without challenge. Indeed, because the scientists could set the quotas down to zero whenever necessary, the scheme effectively established "selective moratoria" on the more endangered species. For instance, all Antarctic fin whales fell under an indefinite moratorium. Furthermore, by dividing all whale species into discrete geographical stocks, the scheme ensured all populations and all regions were covered. In 1976, when the NMP was finally implemented, quotas were effectively reduced for nearly all areas and species. The total allowable catch was down to 72 percent of its 1972 rate. After that the quotas were brought down every year, in accordance with SC recommendations (FoE 1978, 27). An interesting discursive effect, the moratorium proposals that began to proliferate in the late 1970s were technically tabled at the IWC as "zero quotas" in accordance with the language of the NMP. However, during the deliberations they were increasingly referred to as "moratoria," thereby effectively evacuating any reference to "quotas," which implied that whales could still be hunted (when the quota is not set at zero). This subtle shift signaled a new discursive regime taking shape, where whales are not whaled.

"Aboriginal" versus "Commercial" Whaling

The anti-whaling campaign stalled for the next two years over an episode of internal dissent within its strongest ally, the United States. While the United States had outlawed *commercial* whaling in 1969, it still counted whalers among its indigenous populations. The bowhead hunt constituted an ancient Inupiat tradition involving a complicated array of social and religious rites (Hess 1999). The Alaskan Inupiat Eskimos' traditional whaling grounds had been discovered by the New England whalers, severely exploited over the second half of the nineteenth century, then swiftly deserted after 1914 as whales become too scarce. The New Englanders had brought with them technology that made the hunts more efficient, notably guns and motorized vessels. By the 1970s the Alaskan Eskimos had become a politically significant minority, in the wake of the civil rights movement and the rise of indigenous activism. The discovery of important oil and gas reserves on the Alaskan North Slope in the 1960s added to their political clout. The land became the Eskimos' major political card, which they successfully played to ensure permanent access to Washington. In 1971 Congress passed the Alaskan Native Settlement Act, which provided for monetary compensation for the land. The following year the Alaskan settlements organized into the North Slope Borough, a loose federations of villages with powers to levy taxes on the physical structures of exploitation. The bowhead whale had been recognized as highly endangered under the Endangered Species legislation since 1970, and the IWC estimated that it stood at 2 percent of its original population in 1977 (IWC 1978). Yet that same year the U.S. delegation successfully obtained a quota of about ten whales for the Eskimo hunt. This followed on the heels of an Amendment to the ESA successfully passed the year before to provide for a "native exemption" (U.S.C. 1976).[3]

Whaling, which was considered a means of reasserting a culture on the brink of extinction, was revived by the new Alaskan resources: the hunt escalated from twelve whales on average in the 1960s to thirty-two in a decade, in addition to seventy-nine "struck but lost" (Gambell 1995). However, given their state of extreme depletion, IWC scientists soon found that this catch still exerted excessive strain on the stocks. As early as 1972 the SC began requesting additional information on the Alaskan fishery and repeatedly recommended that the Commission urge the United States to reduce its loss rate. The United States ignored these requests. Eventually, at the 1977 annual meeting in Canberra, the SC recommended a zero quota for bowheads—effectively a moratorium—based on NMP calculations. Endorsing these results, the Commission revoked the aboriginal exemption by a majority vote and placed bowheads under total protection.

The United States reacted promptly to the ban proposal. A zero quota would take whale management out of domestic hands altogether. This time, the government initiated comprehensive research into the bowhead stocks, with the hoping of leading to an upward revision of the estimates the SC had been working with. The United States was torn between its whaling and anti-whaling factions, both of whom had powerful lobbies in Washington. Within a few weeks of the IWC meeting, the Inupiat, who had walked out of the IWC meeting, organized into the Alaskan Eskimo Whaling Commission (AEWC) and demanded that the U.S. government file an objection to the zero bowhead quotas, to the greatest alarm of environmental activists. The credibility of the carefully crafted U.S. anti-whaling stance was at stake. The controversy spanned several government departments and newspapers. When the federal government failed to object, the AEWC filed suit in the U.S. District Court, requesting a temporary restraining order to require the federal government to place an objection. The request was granted, only to be promptly appealed by the government. The final court decision swung in favor of the government, to the relief of anti-whaling activists.[4] The government placed no objection but instead took immediate steps to add the bowhead issue to the agenda of a special meeting convened in Tokyo that December to resolve a scientific dispute over a sperm whale quota calculation.

The Tokyo meeting was a technical meeting, removed from the publicity that increasingly surrounded the plenary. By coupling sperm and bowhead quotas, the American Commissioner Richard Frank, soon branded as the "traitor" by the U.S. anti-whaling movement, was able to secure enough votes to obtain a revision of the bowhead quota from zero to "twelve taken or eighteen struck" (whichever occurred first). He supported a revised sperm whale quota to obtain against the votes of sperm whaling countries, the USSR, Australia, and Japan. From the scientists' point of view this trade-off was highly problematic, as it set off a highly endangered species (bowheads) against a nonthreatened species (sperms). For example, the sperm whales hunted by the Japanese (the Northwest Pacific stock) were estimated in the early 1980s at some 200,000, that is, between 50 and 70 percent of initial stock size, which was close to being able establish quotas under the NMP (see chapter 6). This move decisively shifted the criteria for determining quotas from biology to politics.

Whereas the SC was divided on the sperm quotas, on bowheads, they were not. In the face of an unusually homogeneous SC urging for a moratorium on bowhead whaling in Tokyo, the U.S. delegation threw a completely different factor into the equation: the need to take into account "aboriginal cultural and subsistence needs." Taken by surprise, the SC could only acknowledge that this was necessary. Thus the United States won the argument, not by engaging with it on its own grounds, but by displacing the seat of the argumentation, or *topos*. It did so by ushering a completely new category into the decision-making process. This move was entrenched with the passing of a resolution on aboriginal whaling at the next Plenary, which led to the creation of a Special Working Group mandated to develop a distinct aboriginal whaling regime within the IWC. A first Special Meeting on Aboriginal/Subsistence Whaling was convened in Seattle in February 1979; it was followed by a second one in Washington in April to finalize recommendations and "needs-defining factors."

However in July 1979, the SC reasserted its view of the three previous meetings that, "from a biological point of view, the safe course" remained a zero quota, even taking into account the new parameters and revised population estimates (IWC/30/4 1978). At the Plenary, two resolutions were put forward. One, proposed by Australia, seconded by New Zealand, requested a zero quota for bowhead hunt. After its apostasy Australia had become the new official voice of anti-whaling NGOs and (for a time) critic of American whaling. The United States simply retabled the same resolution as the previous year, calling for eighteen landed or twenty-seven struck. Both the Australian and American resolutions failed to garner the necessary votes; yet after more lengthy negotiations, the quotas finally adopted were eighteen landed or twenty-six struck. Hence the only moratorium that had actually been recommended by the IWC's own SC had effectively been thrown out by the United States, despite its being one of the strongest proponents of strengthening the IWC, more particularly its scientific component (M'Gonigle 1980). Simultaneously another motion was passed supporting a National Management Plan proposed by the United States, with which it recovered the responsibility for bowhead management. Authority for determining "cultural needs" was also placed with the U.S. government, which was simply advised to take account of the list of indicators drafted by the IWC working group and to submit an annual report to the IWC. The United States had recovered an authority that had been temporarily lost with the bowhead quota affair. It managed to bypass the SC's advice without technically breaching any rules of procedure.

Of the five moratoria proposed in 1982, the one that was adopted was the U.S. proposal to suspend *commercial* whaling, rather than, for instance, the Australian proposal, which covered all whaling. Furthermore, the United States had successfully won anti-whaling countries to its cause, since of the four other countries that submitted moratorium proposals that year (Australia, Seychelles, United Kingdom, and France), three voted *against* a zero quota proposed by the scientists for the Alaskan bowhead harvest (France abstained). The United States had succeeded in annulling a moratorium strongly recommended by the IWC's own SC while pushing across a moratorium on all other types of whaling *not* recommended by the SC. This is thus an apt example of an exercise in symbolic domination or what, in her analysis of whaling politics, and in a different language altogether, Peterson (1992, 149) captured as "hegemony in the Gramscian sense of winning all others over to its preferred categories for defining and dealing with the problem."[5] The United States had succeeded in imposing upon the international whaling regime a categorization devised to resolve contradictions in its own whaling policies.

This distinction between "aboriginal" and "commercial" whaling has since ordained the entire field of whaling discourses and practices. It has brewed axiomatic articulations whereby the "bad" whaling is "commercial" whereas "aboriginal subsistence whaling" is accepted as an "exotic," "different," "cultural" practice (Freeman 2001). Furthermore, while it may be relevant to Alaskan whaling, the categorization does not reflect the realities of whaling in small coastal villages in Norway and Japan (see chapter 11 for an extensive analysis). Although whalers themselves reject this distinction (World Council of Whalers [WCW] 1998b), they have little say in designing or even using the categories by which whaling is managed today. The Makah of Washington state, for example, were unable to wield that category for their own traditional whaling practices until 1999. Japan for its part has sought to imitate the United States' discursive strategy by coining a new category that reflects the distinctiveness of the whaling that occurs in its four remaining whaling villages, "Small Type Coastal Whaling" (see chapter 11). The power to decide what whaling qualifies as good/aboriginal and bad/commercial whaling—symbolic power—has been taken out of the hands of those who have a material interest in it.

State–NGO Interactions in the Run-Up to the 1982 Moratorium

1979—The First Whale Sanctuary and the Moratorium on Pelagic Whaling The year 1979 marked a turning point in the anti-whaling war. First, the mantle of protection was extending: in the previous year one more important species, the sei whales of the Southern Hemisphere, had come under complete protection under the NMP, and the quota for sperm whales was reduced. Second, the remit of international whaling regulations was expanding. Pressure upon whaling countries to join the IWC had been ratcheting up since 1972, and by 1980 five new whaling countries had joined as a result of American pressures (Spain, Korea, Peru, Chile, and China). Thus strengthened, the IWC was more than ever the stronghold for activists to take over. Third, the ranks of the anti-whaling camp were inflating, with important new allies among former whaling countries, such as Australia (since 1978), the Netherlands (since 1977), and New Zealand (since 1976). In 1979, for the first time, a country that had not been involved in whaling joined the IWC explicitly to buttress the anti-whaling cause, namely, Sweden. This was also the year Sidney Holt officially retired from the FAO.

By 1979 the anti-whaling movement was larger, better funded, more organized and thus more effective than ever. The American Stern fund had hosted a meeting of NGOs in Washington two months ahead of the IWC annual meeting (M'Gonigle 1980). In addition, after the 1978 London meeting a coordinating group had been established in London, the Marine Action Centre, to keep an eye on developments in the IWC and to foster the development of long-term strategies. The Centre inaugurated a monthly newsletter sent out to a worldwide mailing list, the precursor to the e-newsletters still currently used by anti-whaling NGOs. At the IWC Plenary NGOs were also more numerous than ever before: a survey of the *Reports of the International Whaling Commission* (1964–1980) reveals that twenty-one groups attended that year, compared with fifteen in 1978, which was already about twice the habitual NGO attendance. They almost doubled (to forty) the following year (1979). That year, activists came to the IWC better briefed, with well-drafted policy documents to distribute both to the country delegates and the press during the negotiations, thereby inaugurating a practice pursued ever since. The 1979 Plenary was also the first time the IWC opened to the press, unleashing a "media blitzkrieg" upon it (Stoett 1997). As we have seen this was a press already shaped by anti-whaling frames. From then on, it tended to mostly relay the story-lines fed to it by an activist literature made widely available throughout IWC meetings, both in the form of targeted pamphlets and as a collective daily newsletter, *Eco*. FoE, which had successfully organized a 15,000-strong whale rally in London, exceeded 300 press article mentions in July 1979 alone (Stoett 1997, 96). From the onset NGOs thus successfully established themselves as authoritative sources of information for the press on the issue, and to this day, anti-whaling activists are frequently employed as talking heads in the press to explain the IWC annual meetings to the broader public.

However, in 1979 activists also successfully broke into the ranks of the IWC and positioned themselves at key points in the policy-making process. Since 1978 they had become the new watchdogs in the SC as we saw in chapter 6. That year they also infiltrated the Commission itself, where they began drafting the resolutions upon which states voted. The first was a resolution proposed by Panama that mandated the SC to address the question of whales' intelligence rather than their numbers. Their most significant move, however, was to establish themselves as authors not merely of resolutions but of the regulatory instruments themselves. Two pieces of legislation would be at the center of this strategy to recast IWC management as the protection of whales rather than the regulation of whaling: whale sanctuaries, and moratoria proposals. To carry out this strategy they literally took over country delegations, the seat of voting power at the IWC. Fortom-Gouin emerged in 1978 as Commissioner for Panama—the very nation that had harbored Onassis's pirate whaling. Alongside the resolution on cetacean intelligence, he put forward a moratorium proposal. In a dramatic sequel, Panama dismissed Fortom-Gouin before the opening of the Plenary and withdrew the item from the agenda on the first morning.[6] He was, however, able to rejoin the Panamanian delegation as an advisor, where he was instrumental in obtaining the first, highly symbolic partial moratorium on pelagic whaling that brought Antarctic whaling to a halt. This was achieved by introducing a last-minute amendment to the American moratorium, on the table yet again, which separated out the proposal into two distinct moratoria: a coastal one (which was defeated) and a pelagic moratorium (which went through). This was a key victory for the anti-whalers. Panama resigned from the IWC the following year.

The Indian Ocean Whale Sanctuary proposal was introduced in 1979 as part of a twofold initiative put forward by the Seychelles. The first part, a moratorium on sperm whaling, was defeated (notably by the new whaling members), but the sanctuary proposal went through. It carved out the whole Indian Ocean, an important feeding ground for the whales, as a protected area. The proposal was the brainchild of Sidney Holt, who was inspired by a provision of the 1937 Whaling Convention (interview with Sidney Holt; see also Gambell 1997b). According to Holt himself, the move succeeded because it took everyone by surprise. The proposal effectively extended the pelagic moratorium. It established a domestic regulatory instrument for preservation at the international level. Most importantly, it established NGO activists as authoritative conservation strategists. In 1994, Holt was hired by France to carry through a Southern Ocean Whale Sanctuary that it had been proposing for the past two years (see chapter 10). More recently, Brazil has regularly tabled a South Atlantic sanctuary, and Australia and New Zealand one for the South Pacific.[7] As for how the Seychelles came to submit such a proposal, it had joined the IWC in 1979, with, as advisors to the Commissioner, British nationals and environmental activists Lyall Watson and Cornelia Durant.[8] The next two years the advisors were Cornelia Durant again and Sidney Holt himself. Watson had become Alternate Commissioner, and since no Commissioners were nominated both years, he effectively cast the votes for that country.

1980–1982: The Recruitment Drive After the adoption of the first moratorium, and with the new presence of the press at the Plenary, the world's attention was upon the IWC. The next two years witnessed an intense reshuffling of IWC membership. There were successive waves of new entries and some withdrawals. In 1980 Oman and landlocked Switzerland became members. In 1981, Jamaica, St. Lucia, Dominica, Costa

Rica, Uruguay, China, St. Vincent, India, and the Philippines joined. This arrival en masse of new members with little prior involvement in whaling indicated how the debate had begun to ripple beyond the technical issue of whaling management alone. The ranks of the whaling countries, on the other hand, were thinning out. Canada, caught in the throes of the anti-sealing campaign at home, simply resigned from the IWC that year in reaction to domestic environmental activism. China was the only new whaling nation. In 1982 seven more nonwhaling countries joined: Senegal, Kenya, Egypt, Belize, Antigua, Monaco, and Germany. Dominica, who had localized whaling, withdrew, together with Jamaica. In 1983 Finland, another nonwhaler, joined. In less than five years IWC membership had surged from seventeen to thirty-nine nations, and the voting balance decisively tipped against the whaling nations.

This movement to join was not about material payoffs states incurred—there were, for example, few payoffs for Kenya or Finland—but about "fitting in." Indeed by then the IWC had evolved from an obscure club of whalers into a high-profile society of *anti*-whaling states. Rather, two different types of factors help account for this new momentum that drove states to the IWC. First was the indirect, structural effect of a discourse that was becoming entrenched as the dominant global discourse on whales. As such, the anti-whaling discourse staked out the broader field of interactions, within which additional particular interests could then be played out. These included, for example, pandering to the domestic green vote, to which, in the absence of whaling constituencies, anti-whaling policies were a relatively inexpensive way of catering. For example, we saw in chapter 4 how West Germany's accession served the broader purpose of redefining its diplomatic relations against its own history. It was also explicitly a response to a demand by Greenpeace Germany. Key also was the role of IFAW activist Petra Deimer, who served on the delegation and wielded significant influence both over the German Commissioner and in Bonn more generally (Day 1992). She, together with Holt, was also instrumental in obtaining that minke whales be added to Appendix I list of "highly endangered" species at the CITES in 1983, thereby precluding international trade in the species whose endangered status was most contested and that the whaling countries were therefore more likely to hunt (Epstein 2006, Hutton and Dickson 2000, Mofson 2000). Germany's anti-whaling policies have remained an important stance for a reunited Germany and for green party politics.[9]

Second, the involvement of another set of nonwhaling, mainly African, developing countries illustrates how the now dominant anti-whaling discourse was used in another particular play of positioning in relation to the broader society of states. While they added their vote to the moratorium, they were barely involved: some remained in the IWC only a year, others such as Dominica never even appointed a Commissioner, while a few (Egypt or Kenya) merely sent a member of the local embassy to cast the vote on the crucial day (Birnie 1985). They were catering not to their own domestic public opinion but to the perceived expectations of other, more powerful states at a particularly difficult juncture in the international political economy of North–South relations. The early 1970s had been hopeful times for developing countries, who had proclaimed their new majority in the UN under New Economic International Order. The 1980s by contrast had inaugurated a new era of aggressive liberalism with the Reagan–Thatcher dyad. Aid fatigue progressively settled in, and the developing countries lost their negotiating advantage (Persaud 2001).[10] The 1982 moratorium vote provided a good public relations gesture vis-à-vis developed countries preoccupied with whales, in the context, furthermore, of the final negotiations of UNCLOS III, which was opened for signature that same year (Stoett 1997).

While this was clearly instrumental behavior on behalf of both Germany and these African countries, my main point here is that the ability to pursue such instrumental goals is itself contingent upon the existence of a prior discourse that maps out these goals as worth pursuing. In other words, the fact that these were instrumental actions (one in relation to domestic politics and the other in diplomatic North–South relations) does not take away from the fact that there existed a constitutive discourse within which these states had to position themselves in order to be able to *be* instrumental. Without this discourse, it would not have made any sense to be nice to the whales. There was, in other words, no material instrumentality preexisting the discourse itself.

Alongside these indirect, systemic effects a second type of more direct pressure was deployed by activists upon country delegations at the IWC. We have already seen with Seychelles and Panama how activists began infiltrating the ranks not only of their own countries but of third-country delegations. In the run-up to the 1982 moratorium this developed into a full-fledged strategy. In the words of investigative journalist Leslie Spencer (1991):

... Membership in the Commission was open to any country that was willing to pay an annual fee of roughly $20,000 to $30,000 plus the cost of sending its representative to meetings. According to Francisco Palacio, a former Greenpeace consultant on marine mammals, he and McTaggart, working with their friends, came up with a way to bend the commission to the Greenpeace view that there should be an outright ban on whaling.

The whale savers targeted poor nations plus some small, newly independent ones like Antigua and St. Lucia. They drafted the required membership documents for submission to the U.S. State Department. They assigned themselves or their friends as the scientists and commissioners to represent these nations at the whaling commission.

The recruitment drive had been launched. The following is drawn from personal correspondence from a disillusioned NGO official.[11] The Indian Ocean Sanctuary Account was set up to channel the funds from five of the IFAW country offices, which were thus not traceable to a single payment and remained unaccounted for within the general IFAW accounts. It was used to pay the Seychelles' membership fees and all travel expenses of its Commissioner, who was not a national from that country.[12] The accounts were in the hands of Sidney Holt, whose expenses were also covered by WWF (Pearce 1991, 24; Stoett 1997, 94). Additional finances were provided by Fortom-Gouin, which the NGO official discovered in 1987 to be of illegal origin.

Between 1979 and 1985, the IWC Plenary became a grand activist ballet of musical chairs. St. Lucia, a sovereign nation since 1979, joined in 1981 and took over from Panama's role as "spokesnation" for the anti-whalers. Next to St. Lucia's Commissioner stood Palacio, officially the Alternate Commissioner, who was *also* registered as Commissioner for Columbia, then attending as an observer state. Besides him in turn on the Colombian delegation stood the ubiquitous Fortom-Gouin. Both of them appeared on the St. Lucia delegation the following year (1982), Palacio as Alternate Commissioner.[13] Across the table that year his lawyer, the American Richard Baron, was Commissioner for the newly joined Antigua and Barbuda (1982), who had secured independence from the United Kingdom only the year before.[14] Further down the table Lyall Watson was that year Alternate Commissioner for Seychelles, advised by Durrant and Holt.[15] St. Vincent and the Grenadines (independent only since 1979) also joined that year with, as Commissioner, the American Christopher Davey, Chairman of the Cetuman Foundation and anaesthetist in Miami, who equally featured as Commissioner for the observer state Costa Rica.[16] Costa Rica deposited its instrument of accession just after the end of the meeting, establishing it as a member for 1981. Again, quoting Spencer (1991):

> According to Palacio, the Greenpeace-inspired commissioners enjoyed an annual all-expenses-paid ten-day trip with a $300-per-diem perk to attend commission meetings. Palacio says the group paid to fly a U.N. ambassador home to talk his government into going along with the plan.
>
> Between 1978 and 1982, Palacio says, the operation added at least half a dozen new member countries to the commission's membership to achieve the three-fourths majority necessary for a moratorium on commercial whaling, which passed in 1982.
>
> This project cost millions, says Palacio, including the commission membership payments picked up on behalf of cooperating members. "In membership fees the payments amounted to about $150,000 [a year], and then we had all the grease money throughout the years" says Palacio. The Frenchman Gouin, then in his 30s, was the angel, funneling the funds through a Miami-based "foundation" called the Sea Life Resources Institute. Where did Gouin get that kind of money? From trading investments, he says ...

D-Day The four years between 1978 and 1982 saw a large variety of moratorium proposals being put forward by different countries as they joined the anti-whaling ranks. Whereas previously alliances still fluctuated according to the species under consideration, positions were increasingly hardened, and the IWC split between a shrinking whaling and a growing anti-whaling camp. A few countries still attempted to occupy the middle ground, such as Denmark, for one, a European Community member where whaling still occurred (in Greenland and the Faeroes Islands), or Argentina, which was increasingly uneasy at the SC having to examine issues such as whale sanctuaries. By the same token, both sanctuaries and moratoria were becoming entrenched as key regulatory tools of environmental activism-cum-policy-making.[17] Meanwhile the pressure on the pavement continued unabated: in the run-up to the 1982 annual conference in Brighton, the FoE rally had grown to 20,000 demonstrators (plus Willy the rubber Whale). That year, five moratorium proposals (from Seychelles, France, Australia, the United Kingdom, and the United States) were tabled. Finally, an over three-quarter majority vote (twenty-five) entrenched a ten-year moratorium on commercial whaling to take effect in 1985 after a three-year phasing out period. Seven countries had voted against it, the USSR, Japan, Peru, Iceland, Norway, Brazil, and Korea; Denmark and South Africa abstained. Popping champagne corks in the negotiating forum echoed crowd cheers outside. Under the "Save the Whales" banner another was added, "Whales Saved—Brighton 1982" (Day 1992).

Conclusion

This chapter retraced the progression of the anti-whaling campaign. In under two decades, whaling had been successfully denormalized through the relentless efforts of a set of disparate and dedicated activists who rallied under a new discourse on whales. They propagated this discourse from the ground up, by targeting two sets of actors successively: first, individual members of an increasingly global public, and then states. As it took hold, the discourse thus progressively enfolded state delegations and yielded a new, global anti-whaling discourse coalition of states and NGOs. This discourse coalition, which is still in place today, tipped the balance of power between the two discourses. It establishing the anti-whaling discourse as the dominant discourse on whales and whaling and the pro-whaling discourse as the new "discourse of resistance," a point that will be further developed in chapters 10 and 11.

The suspension of commercial whaling worldwide in 1982 was the last act in the transformation from a "society of whaling states" to a "society of anti-whaling states" that centered upon the IWC and extended to other international forums as well, such as CITES, UNEP, and the FAO as we saw in chapters 5 and 6. It was an unequivocal victory for activists, who achieved it indirectly, by wielding the new influence upon states afforded them by the rise to prominence of their discourse, but also directly, by taking over state delegations at the IWC. This is not reducible to NGOs' being better funded and better equipped than those countries. These states' espousing anti-whaling policies was instrumental both for them and for the NGOs who convinced them to do so. However, what made possible this marriage of convenience was a prior discourse that made it a good idea to be wed in the first place. In other words, there is no such a thing as a priori instrumentalism. To consider the discourse is effectively to consider the conditions of possibility for such instrumentalism, and how it, in turn, made possible specific actions.

This account is useful for fleshing out a distinction that has been running throughout some of the previous chapter: that between *the actor* (state and individual), on the one hand, and *the subject* (I/we) of a discourse. The latter constitutes a particular place carved out within a discourse, into which an actor steps to take up that specific subject-position—to speak as the subject (we/I) of that discourse. Thus what a discourse can produce is a subject-position, not a (flesh-and-bone) actor. Moreover, the anti-whaling campaign illustrates how one same actor can take on several different, even contradictory, I/we(s), depending on the specific sphere of action in which it is intervening. For this campaign featured several (flesh-and-bone) actors speaking as subjects of several different discourses: as activists, scientists, even as states.

III

Reproducing the Anti-Whaling Order

8

Crafting the Anti-Whaler (I): An Applied Discourse Analysis

Part II focused on the production of a new dominant global discourse about whales entrenched by the 1982 IWC moratorium on commercial whaling. Part III will examine the factors that have enabled its reproduction. This discourse has been able to last such that, although it was adopted as a ten-year moratorium, it is still firmly in place today. This chapter argues that the anti-whaling discourse created a particular subjectivity, that of the anti-whaler. It uses discourse analysis to appraise the making of an anti-whaling subjectivity. Indeed, the creation of an anti-whaling *identity* is what explains the purchase of the anti-whaling discourse—its attractiveness, its longevity, in other words, its power. Otherwise, the volume of talk about whales would be far more muted. Whaling may no doubt have come to an end eventually because of economic factors alone (see chapter 2). However, the important point is that, even when it did, whales continued to matter so much, because, I argue, they were involved in the creation of a new, green, identity.

The identity proposed by the anti-whaling discourse is foremost an individual identity. As we saw in chapter 7, the anti-whaling discourse addressed primarily individuals (the members of a globalizing media audience), inviting them to join in the battle to save the whales all around the world. Hence the analysis here is scaled down, in this and the following chapter (chapter 9), to the level of the individual, which is the level at which the notion of "subjectivity" properly applies (Butler 1997). Centrally, however, this was an identity that *also* mattered for states. For once their citizens spoke the anti-whaling discourse, certain courses of action became precluded for a state (such as promoting whaling, see also Shaw 2003, 2002; Connolly 1989). The way in which states, in turn, have positioned themselves in relation to this subjectivity will be examined in chapter 10. Discourse analysis is thus the appropriate method for studying a type of power that works from the ground up, the power of discourse. Before proceeding to apply it, however, I briefly recall and expand upon distinctions introduced in the opening of this book, namely, between "subject" and "subjectivity," and how these, in turn, relate to "identity."

The analysis in part II foregrounded "the subject" as a discursive category, that is, as a position with a discourse. Chapter 5 in particular focused on the ways in which the anti-whaling discourse produced a powerful subject-position from which particular actors (individuals or states) can speak and be heard. What was left suspended with such a focus was the ways in which discursive subject-positions are bound up with particular *identities*. This dimension is developed here by considering the relationship between "subject" and "subjectivity." First of all, the *subject* is, for the discourse theorist, a place in a discourse, that is, before actual texts to be examined, a grammatical category (the subject of a sentence). The term thus applies to the actual analysis of discourses. Instead *subjectivity* refers to the identity taken on by flesh-and-bone individuals. They capture two sides of the coin, shining two different analytical lights upon the same phenomenon, namely, the construction of an identity.

Carrying these insights into the method of discourse analysis, the subject is the grammatical category most relevant in examining the construction of subjectivities. In turn, *predicate analysis* is the type of analysis most suited to examining the discursive formation of a particular subjectivity (see also Milliken 1999). It is a means of unpacking particular *articulations* (see chapter 5) brought into play by a discourse. A "predicate," in the logician's and the linguist's language, is a quality or an attribute. Put simply, predication captures the way in which "something is said" about the subject of a sentence. A sentence, at its most elementary, is constituted by a verb linking a predicate to a subject. That is, the subject is attributed with particular qualities or capacities. It is characterized as *being* something, *having* something, or *doing* something, all of which constitute types of predication.[1] Patterns of predication are especially important in considering the formation of particular subject-positions in a discourse.

Speaking the anti-whaling discourse marks you as someone who cares about whales. In other words, it does not just do something for the whales (according to the discourse itself) but it says something about the individual who has stepped into that subject-position. It casts her as a compassionate, environmentally mindful, and quintessentially "good" person. First, this "stepping into" is envisaged here as a dynamic process. That is, identity is not a fixed, static, and preimposed category. Rather, it is an active process of *identification.* An identity requires an individual actively embracing it. It demands active recognition on behalf of the individual. Hence it must be something she can identify with. "Being anti-whaling" must be something she wants to be, and be seen as being, by her peers. Thus of key concern in analyzing anti-whaling texts are the rhetorical strategies by which the individual is drawn into this subject-position, or *interpellated* (see chapter 5).

That an individual steps into the anti whaling subject-position does not mean that she or he is *reducible* to that identity. An individual may bear several different identities, some indeed that she may consider more fundamental than "being anti-whaling." An individual may also change identities, and we saw ample examples of this in the last two chapters (with the figure of Sidney Holt, for example). Nonetheless, the moment that individual is speaking (or writing, or chanting, or marching along with) the anti-whaling discourse, she is positioning herself as an anti-whaler. In that particular moment she is marking herself as an anti-whaler, even if that may not matter as much as some of the other subjectivities she considers as hers, and even if it is shed off, say, once the whale march is over. Moreover, the anti-whaling campaign described in chapter 7 has shown that there are very dedicated activists for whom this subjectivity *was* in fact central to their self-definition, insofar as it shaped both their professional careers and how they saw themselves on a more personal level. In other words, this is about taking discourses seriously as the locus of production and reproduction of identities. You are who you say you are, or rather you are who you speak *as.* And that you may pretend to be someone else is irrelevant in this perspective, for you are still defining yourself as that particular person when speaking that discourse.[2] Most importantly, in so doing, you are engaging with the particular relations of power that are played out around that discourse.

Choice of Texts

Methodologically, a text offers a snapshot of the broader discourse under examination, capturing its salient features at a particular point in time. The text analyzed here is a 1974 boycott advertisement by a coalition of U.S. environmental groups published simultaneously in several major national newspapers (see figure 8.1). This particular sample is

SAVE THE WHALES!

鯨を救おう

THEY ARE BEING SLAUGHTERED TO EXTINCTION
BY JAPANESE AND SOVIET WHALERS

BOYCOTT JAPANESE GOODS

SUCH AS CARS, MOTORCYCLES, TELEVISIONS, STEREOS, CAMERAS

DON'T FLY JAPANESE AIR LINES
DON'T BUY RUSSIAN FURS AND VODKA

THE JAPANESE ARE THE WORLD'S BIGGEST WHALE KILLERS. The whaling companies are owned and controlled by the great manufacturing, trading and financial combines that produce and market the vast array of Japanese products sold in America. For example, Nissan Motor Co., which makes Datsun automobiles, holds major interest in Nippon Suisan, the world's largest whaling company, through the auto-maker's insurance affiliate, Nissan Fire and Marine Insurance, and through Industrial Bank of Japan. Other Japanese conglomerates with substantial direct or indirect interest in the whaling companies include Mitsubishi, Mitsui, Marubeni, Nippon Steel, Hitachi Shipbuilding, Nomura Securities, Tokio Marine & Fire Insurance, Daiwa Bank, Long-Term Credit Bank of Japan, Sanwa Bank, Fuji Bank, Dai-Ichi Kangyo Bank, and Sony.

The Japanese business community opposes all international attempts to halt the slaughter that is driving the great whales to extinction. Likewise, industrial and military interests in the Soviet Union continue the whale slaughter despite warnings from Soviet scientists. The Russians are the world's second biggest whale-killers.

YOU can help save the whales by joining the boycott against the whale-killers.

More than 2,000,000 whales have been killed in the past 50 years, so many that all eight species of great whales are in danger of extinction. The Blue Whale, the largest creature ever to live on earth, survives in such few numbers that it may be past the point of saving.

The United Nations has three times unanimously called for a moratorium on whaling. At the U.N. Conference on the Human Environment at Stockholm in 1972, the international community voted 53 to 0 for a halt to whaling. At two subsequent conferences of the U.N. Environment Program, in Geneva and Nairobi, there were unanimous calls to stop the slaughter.

The Japanese, and the Soviets, have arrogantly ignored the international pleas.

Japan and the Soviet Union have consistently opposed moves by the International Whaling Commission to save the whales from commercial extinction. At the 1973 meeting of the IWC, lower and more stringent quotas were established on fin, sperm and minke whales. Both Japan and the Soviet Union angrily opposed the rulings. They announced they would defy the lower quotas. Thousands of whales died last winter because of this defiance.

At the 1974 IWC meeting, the Japanese again bitterly opposed all attempts to halt or even reduce the whale slaughter. Thirteen member nations approved a selective moratorium on the most endangered species, to begin in 1976. But Japan, joined by the Soviet Union, arrogantly voted no. According to reports from the Japanese government, the whalers will seek to nullify the selective moratorium before 1976. Indeed, the Scientific Committee of the IWC is meeting in La Jolla this week to develop guidelines for the selective moratorium.

The slaughter continues relentlessly. Between them, Japan and the Soviet Union kill over 80% of the whales. Every 14 minutes another of these gentle, intelligent animals is harpooned to agonizing death.

In response to this outrage, every major American environmental and animal protection group has joined to call for a boycott of Japanese and Soviet goods. Until the whale slaughter is stopped, the boycott campaign will be pursued and publicized in all Japanese and Soviet markets.

THE BOYCOTT CAMPAIGN IS
SUPPORTED BY THESE CONSERVATION GROUPS:

National Audubon Society	Sierra Club
Friends of the Earth	National Wildlife Federation
Environmental Defense Fund	The Fund for Animals
Humane Society of the United States	The Wilderness Society
National Parks and Conservation Assoc.	Rare Animal Relief Effort
Center for Science in the Public Interest	Environmental Action
Explorers Club Environment Committee	Defenders of Wildlife
Society for Animal Protective Legislation	Environmental Policy Center
New York Zoological Society	Animal Welfare Institute

In an attempt to undermine the Save the Whales campaign, some Japanese interests have sought to label the boycott as racist. It is not. The effort is aimed at the corporations and governments of Japan and the Soviet Union which refuse to heed opinion of scientists and the international community. The boycott was undertaken only after the whalers had stymied every effort to bring reason and responsibility to whaling. The Japanese-American Citizens League, representing hundreds of thousands of Americans of Japanese ancestry, has condemned the notoriety generated by the Japanese whalers. At the JACL convention in July, the League unanimously supported a worldwide moratorium on whaling.

Figure 8.1
A sample anti-whaling text (I). Source: Scanned from *Los Angeles Times* 06/12/74, viii, 6.

The United States has placed all eight of the great whales on the Endangered Species List. The government has banned commercial whaling and the importation of all whale products. Secretary of State Henry Kissinger last fall sharply criticized Japan for ignoring international controls on whaling. Said Kissinger in a cable to the Japanese government: "The United States Government is extremely disappointed with the backward step occasioned by Japanese objections and it is hopeful the Japanese Government will reconsider."

JAPANESE HYPOCRISY

The Japanese claim that whale meat is vital to the Japanese diet. In fact, it provides just 1% of the protein of the Japanese people. The whaling companies are guilty of utter hypocrisy when they make this claim. The same whaling companies also own the world's largest fishing fleets, and they export enormous quantities of fish products—protein that can easily replace whale meat.

Modern whaling is a savage, ruthless exercise, nothing like the romantic days of 19th-century whaling. Here is an eyewitness description of modern whaling by Dr. Harry Lillie, who sailed as a ship's surgeon:

"The present-day hunting harpoon is a horrible 150-pound weapon carrying an explosive head which bursts generally in the whale's intestines, and the sight of one of these creatures pouring blood and gasping along on the surface towing a 400-ton catching vessel by a heavy harpoon rope, is pitiful. So often an hour or more of torture is inflicted before the agony ends in death. I have experienced a case of five hours and nine harpoons needed to kill one mother Blue Whale.

"If we could imagine a horse having two or three explosive spears driven into it, and then made to drag a heavy butcher's truck while blood poured over the roadway until the animal collapsed an hour or more later, we should have some idea of what a whale goes through."

These great, defenseless creatures are hunted down and slaughtered by sophisticated military weapons. The huge Japanese and Soviet whaling fleets decimate whole herds of whales each day. The whales are sought out by airplanes and helicopters, then are tracked relentlessly by sonar and radar. The terror-stricken whales are chased down by high-speed catcher boats. They are blown up in agonizing death by grenade-tipped harpoons. Huge factory ships rapidly process the whales into commercial "products."

The more than 30,000 whales killed this year by the Japanese and Soviets ended up as pet food, lubricating oil, fertilizer, soap, cosmetics, even delicacies on the Japanese table. There is no need for any of these products in the world economy. There are cheap, ready substitutes for them all. The Soviets feed much of the whale meat to minks and sables, which are then sold to the West for cold cash. Perhaps the ultimate irony of the whale tragedy is that thousands of sperm whales are victims of the nuclear arms race. The Soviet military machine uses sperm oil to lubricate its giant intercontinental missiles.

It is pure greed that drives the whalers. They are cynically slaughtering the whales to extinction. The Japanese and Soviets claim they are preserving the whale populations. But this argument is demolished by a 1974 report of the U.N. Food and Agriculture Organization prepared in cooperation with the Scientific Committee of the International Whaling Commission. States the report:

"It is not realistic to presume (the whalers) will ensure either the maintenance or even the continued existence of a whale stock." Because the remaining whales are being killed at a rate higher than replacement levels, the report concluded, "It pays to exploit a resource (the whales) excessively heavily, even to extinction, and not on a sustainable basis."

There is no doubt the whales are rapidly disappearing. Catch Per Unit of Effort, the method universally used by the fishing industry to determine resource stocks, is falling steadily despite all the new technologies of whale killing.

SAYONARA WHALES

This week the Japanese and Soviet whaling armadas—factory ships, tankers and scores of catcher boats—set sail for the Antarctic, the last refuge of the whales, bringing more death to the dying species.

Only the California gray whales, now beginning their annual migration down the coast from Alaska to the warm lagoons of Baja California, are safe from the deadly harpoons. They are protected by a special international treaty. At the brink of extinction 36 years ago, the California gray whales are now flourishing. The greedy Japanese whalers have been proposing renewed "harvesting" of our local whales, but the suggestions have been strongly rebuffed by the governments of the United States and Mexico. Once there were thousands of Korean gray whales, cousins of the California species, but they were totally wiped out by the Japanese and Soviets.

The fate of the great whales was foreseen more than a century ago by Herman Melville, who sailed the Pacific aboard a whaler. The author of the whaling classic "Moby Dick" wrote: "The moot point is, whether Leviathan can long endure so wide a chase, and so remorseless a havoc; whether he must not at last be exterminated from the waters, and the last whale, like the last man, smoke his last pipe, and then himself evaporate in the final puff."

Jacques Cousteau, French oceanographer, writes: "Our century, as it turned out, was much more destructive to the whales than the period of the great romantic hunts as described in the work of Herman Melville. A hundred years ago, a whaler's three-year expedition netted him 37 whales. Today, a whaler's modern weapons and fast boats give him one whale a day, and sometimes three or four.

"The only creatures on earth that have bigger—and maybe better—brains than humans are the Cetacea, the whales and dolphins. Perhaps they could one day tell us something important, but it is unlikely that we will hear it. Because we are coldly, efficiently and economically killing them off."

THE WHALES NEED YOUR HELP

Please support the boycott. If a product says MADE IN JAPAN or MADE IN U.S.S.R., don't buy it. Purchase goods made elsewhere. Tell merchants WHY you are boycotting Japanese and Russian products and urge them to stock goods made in other countries.

Tell your friends about the plight of the whales and how they can help. The more people who act NOW, the sooner we can halt the needless slaughter that is driving the great whales to extinction. It is time we call a moratorium on buying Japanese and Russian goods until they agree to a moratorium on whaling.

You can help save the whales and publicize the boycott campaign by making a TAX-DEDUCTIBLE CONTRIBUTION to the Save the Whales Campaign. If you believe in this effort, please give generously.

PLEASE SEND ME MORE INFORMATION ABOUT WHALES AND HOW I CAN HELP PREVENT THEIR EXTINCTION.

Name ______________________

Address ______________________

City ______________ State ______________ Zip ______________

Enclosed please find $__________ to help the campaign to save the whales.

Please send me ☐ bumper sticker ☐ button ☐ poster ☐ petition. L.A.

(Make checks payable to the Animal Welfare Institute, which is coordinating the campaign.)

ANIMAL WELFARE INSTITUTE P.O. Box 3650 Washington, D.C. 20007 | Christine Stevens, President

Figure 8.1
(continued)

drawn from the June 12, 1974, *LA Times*. The whaling moratorium proposal, which had first surfaced at the 1972 UN environment conference, had been defeated at the 1973 IWC annual meeting, and that year's IWC annual meeting was scheduled for a few weeks later. The year 1974 also marked the convergence of activist and government actions, as the Pelly Amendment was used for the first time (see chapter 7). This text was chosen because of its position on the trajectory of the anti-whaling campaign, to provide a snapshot into a dominant discourse in the making. It proved very effective, in that it successfully launched one of the first anti-whaling consumer boycotts targeting entire countries (Japan and the USSR) rather than an industry. In fact, it inaugurated the resort to country boycotts as a key campaigning strategy for the anti-whaling campaign that continues to be employed today. It thus provides a good illustration of how discourse is mobilized within a specific NGO strategy to effect change, and speaks to the relationship between activism and discourse. However, from a discourse analytic perspective the point of interest about this boycott text is not that the boycott was actually effective but that it established the call for boycott as a recurrent feature of the anti-whaling discourse, as evidenced by the second text, the "Mad Whalers Disease" caricature (see figure 8.2).[3]

Figure 8.2
An anti-whaling text (II). Source: http://www.nomoredolphins.zoomshare.com/; accessed July 30, 2007.

Text selection raises the broader question of the relationship between *text* and *discourse*, and the methodological issue of whether the chosen text accurately represents the discourse it is taken to express. This in turn requires addressing the question of what constitutes adequate criteria for validity. *Discourse analysis* is distinct from other forms of quantitative, text-based analyses, notably *content analysis*; insofar as analyzing meaning—the core concern of discourse analysis—cannot be derived from the number of occurrences of a locution either in one text or across a sample of coded texts.[4] It requires a more fundamental shift to an *interpretative* mode of analysis. That is, it requires assuming the subject-position that the discourse presumes and reading critically from there. Thus the real validation test for discourse analysis is whether the reading of a particular text has yielded salient categories, that is, categories that continue to illuminate the play of significations conveyed in ulterior instantiations of, in this case, the anti-whaling discourse. As Jennifer Milliken (1999, 234) puts its, "an analysis can be said to be complete (validated) when upon adding new texts and comparing their object spaces, the researcher finds consistently that the theoretical categories she has generated work for those texts." In the case of dominant discourses, an additional test is the test of time, insofar as a dominant discourse is one that lasts; hence its categories are enduringly reproduced. With this in mind, the analysis here is structured as follows: it is centered on the original 1974 boycott text, and parallels are drawn with a more recent call for boycotts, the "Mad Whalers Disease" caricature that circulated on the Internet at the time when this book went to print.

This chapter thus employs the method of discourse analysis to examine the processes of identity construction involved in the anti-whaling discourse as revealed by an anti-whaling text. The anti-whaling identity is constituted around a neat cleavage between "us" and "them." Hence the analysis observes the specific processes of "othering" that serve to constitute the anti-whaling "self." It unpacks, first, the ways in which a "them-whalers" is drawn up, in opposition to which, second, an "us-anti-whalers" is established.

The Construction of "Them-Whalers"

The first question addressed to the text concerns the allocation of *agency:* who is constituted as active; who is passive; who can do what? However, the rhetorical issue of agency opens up the moral issue of responsibility.

The way agency is allocated in the text reflects the way the story-line attributes responsibility for the plight of the whales and, from there, the way it stakes out possibilities for action (see figure 8.1). Agency is allocated in the choice of subjects, objects, and verbs. Positing whales as the passive subjects of a radically new discourse was the founding move of the anti-whaling discourse at large. It is reflected in this text, whose key rhetorical strategy is to cast the whales as entirely powerless. Two lexicons are blended together in portraying the whales: that of victimization, evoked with graphic images (such as "terror-stricken," line 59, "great, defenseless creatures," line 57, "pouring blood and gasping," line 51), and endangeredness ("in danger of extinction," lines 12–13, "survives in such few numbers," line 13), an established, powerful trope in popular discourses at the time, as we have seen.

The figure of the synecdoche features once again as an important trope in the construction of the passive subject (see chapter 5), although it is deployed slightly differently here: to collapse the species to the genus, and thus reduce the many different types of whales to a generic "the whales." The creation of the figure of the "Superwhale" (Kalland 1994b, 1994c) is crucial both to the cogency and the durability of the anti-whaling discourse. It forecloses the possibility of distinguishing between the various types of whales: those that are "cruel killers" themselves/those that are not; those that are endangered/those that are not; those that display sociable behavior comparable to humans/those that do not. For allowing for such distinctions to take shape would undermine a posture camped as "anti-all whaling," and therefore *for all* whales. The positing of a generic "the whales" also serves to adjudicate on the question of whales' endangeredness and perform away any lingering scientific doubt. Thus despite the rejection two years earlier by IWC scientists of the proposal for a blanket whaling moratorium, on the grounds of persisting scientific uncertainty and evidence of the wide divergence in the status of whale stocks (see chapter 7), the text unequivocally asserts that "there is no doubt the whales are rapidly disappearing" (line 74). Because the anti-whaling discourse thrives on the whales' proclaimed endangeredness, one persisting tension it faces in the long run is how to accommodate evidence of whale stock recovery, which, paradoxically, it also *needs* to be able to celebrate its own success and promote alternative whale-related activities such as whale-watching (see the following chapter). This tension already surfaces in this early text, with the reference to recovering whale stocks of California gray whales (lines 78–80).[5]

This first passive subject, in turn, conjures up two radically opposite, *active* subjects. First, the elaboration of a potent, evil agency is central to the text's rhetorical strategy, insofar as the call for action is commensurate with the force of this agency: the more harm they inflict upon whales, the more pressing the need to react. The verbs associated with the whalers are thus strong action verbs pertaining to two narrow lexical fields, that of defiant obstruction ("opposed" lines 19 and 21, "voted no," line 25), and inhumane killing ("tracked," line 59, "hunted down and slaughtered," line 57, "decimate[d]" line 58), doubled with adverbs–markers of intensity ("relentlessly," lines 59 and 28, "angrily," line 21). Second, carving out a first "bad" agent also serves to designate, by implication, a place for the "good" agent into which the reader/viewer is *interpellated* or invited to step, that of the anti-whaling activist. The distribution of agency thus hinges upon a straightforward opposition between "them-whalers" and "us-anti-whaling activists," according to the basic moral scheme: what they do to whales and what we can do to stop them.

The Good, the Bad, and the Whales

Agency is thus shared between a bad and a good agent. The evil agent ("them") compels the righteous agent ("us") to take action. Who, then, are "they" against whom "we" are summoned to act? They are, of course, the "whale killers" (note the chiasmic reversal from "killer whales"). They take shape as entity both carefully bounded without and hierarchized within; although this hierarchy only progressively transpires. Thus while the headlines target "Japanese and Soviet whalers," the first line shifts to "The Japanese." The text itself constantly oscillates between reference to "the Japanese" in general (lines 1 and 44), or "the Japanese business community" (line 9), and the whalers, thereby implying that whale killing is a national trait. The Japanese are established as primarily responsible for the slaughter of whales and thus as prime targets of anti-whaling actions. Consider the phrase "but Japan, joined by the Soviet Union, arrogantly voted no" (lines 24–25). This ranking of targets seems practical enough, given that the volume of trade with Japan was higher than that with the "second biggest whale-killers" (line 10), the USSR. However, it ran counter to whaling facts, given that, in 1973–1974, the USSR was killing more whales than Japan (Schneider and Pearce 2004, 546). Thus more than the realities of whaling, the text's broader political–economic context is relevant to capturing the particular strategies of othering that were being brought into play in the production of this hierarchy. For the run-up to the 1982 moratorium—the years in which the anti-whaling discourse flourished—coincided with a second decade of economic takeoff for Japan, which, from a U.S.-administered, vanquished country was rapidly turning into a full-fledged economic rival. Thus the extensive listing of the Japanese companies (lines 3–7), specifically automobile firms (Nissan Motor Co.), which stand to profit from the slaughter of the whales at the opening of the call for boycott is significant. By the 1980s, Japanese exports to the United States constituted almost half of the latter's annual trade deficit. David Campbell (1994) has shown how, although Japanese investment never actually exceeded 24 percent of net foreign direct investment in the early 1980s, and the bulk of foreign assets were actually held by the British and the Dutch, perceptions of a "Japanese threat" served particular productions of an American "Self" against a Japanese "Other" in U.S. national security discourses. He remarks (1994, 152) how "Japanese economic power was popularly considered a greater national security threat than Soviet military power."

The "Japan problem" had become an official concern under the Nixon administration, which had launched in 1973 the Tokyo round of the General Agreement on Tariffs and Trade (GATT) and brought intense pressure to bear on Japan for its protectionist policies (Francks, Boestel, and Kim 1999). By contrast, Nixon was also the first U.S. president to visit Soviet Moscow in 1972 and initiate the détente in U.S.–USSR military and economic relations. Hence the design of the boycott mirrored the shifting hierarchy of U.S. foreign threats. Or rather, to establish Japan as the worst offender, the text conflates two forms of contemporary dangers: the fear of an economic invasion and the Cold War military threat. Notwithstanding the fact that the former was then a demilitarized nation, the text lumps together Japan and USSR as dangerous military powers, highlighting the "sophisticated military weapons" used by the "huge Japanese and Soviet whaling fleets" (lines 57–58). As for sperm whales, the "Soviet military machine" (line 66) alone is responsible for their decimation, since it "uses sperm oil to lubricate its giant intercontinental missiles" (line 66). Eluded entirely is the United States' own military machine, which used the same oil (see chapter 3), or the fact that it constituted the other main protagonist in the very "nuclear arms race" (line 66) foregrounded here as a main driver of whale extermination.

This omission points to the need to consider who was *not* included in the construction of this whaling other. First, the text draws a line between the Japanese–Americans and those "other" Japanese, roping the Japanese–American Citizens League (lines 36–38) in on the "right" side. Of course, the extent to which the supporting statement of the Japanese–American Citizens League may have been motivated by the desire for acceptance or the recent history of Japanese–American relations in a post-World War II context is not a consideration for this text. That they wrote a statement, however, is also an indication that the Japanese–American community was having to position itself in relation to what was emerging as the only acceptable stance—the *doxa*—on whaling. This indicates the *symbolic power* that was gathering around the anti-whaling discourse, its newfound power to command willful participation. Here the anti-whaling discourse explicitly proclaims itself as not "racist" (line 34). In fact, further undoing any potential allegation, "racist" becomes merely a label abused by the Japanese, in their all-potent agency, to try to undermine this rightful boycott (lines 33–34).[6]

It is all the more significant in this regard that the line drawn around "them-whalers" stops at the whaling nation that is most "like us," namely, Norway, given both Norway's historic whaling role and the fact that it was still then the third largest whaling nation. Moreover, the complete omission of any reference to the responsibility of Western whaling nations in the decimation of certain whale species is striking, insofar as the text itself invokes the history of whaling, flagging the "2,000,000 whales … killed in the past 50 years" (line 12; see chapter 3). Not only is the historic role of the United States as first whaling nation completely evacuated, including in the decimation of the very species that the text itself brands as closest to extinction (the California gray whale), but the U.S. government is actively portrayed as a manner of superagent leading the fight against cetacean extermination. It has listed the whales on the Endangered Species list, "banned commercial whaling and the importation of … whale products" (lines 39–40). Nor is any mention made of the role of Western nations as consumers of whale products, including whale meat—the fact, for example, that some two decades earlier, whale meat was consumed in British schools. In sum, the active denial of Western involvement in whaling at large is a linchpin of the anti-whaling discourse.

Racializing the Other

The real root of the menace is not actually what "they" *have* (the size of their fleets) or even or how they *act* (their agency); it has to do with *who they are*. It is an identity claim. It has to do with the traits inherent to their nationality, their "Japaneseness" and "Sovietness," that is, their essences. Indeed (line 43), "JAPANESE HYPOCRISY," thus highlighted by the typesetting, is what the whales really have to fear. For example, consider the text's patterns of predication, in order to examine more closely how the evil subject—the whalers—are construed. Predication, specifically in the use of adjectives and adverbs, plays a key role in setting up a ruthless ("relentlessly," line 28), angry (line 21), "bitterly" (line 23) defiant (line 22), arrogant (line 25), and cynical (line 67) "Japanese and Soviets." It is structured around a series of binary oppositions that fall on either side of the central axis "us versus them." The first is "rational versus irrational." The registers of reasonableness and responsibility are carefully ascribed to the anti-whaling endeavor. It is not only morally right but it is profoundly rational. By contrast, the Japanese and Soviets are irate and irrational, what's more, systematically having "consistently opposed" (line 19) all efforts to "bring reason and responsibility" (line 36). The portrait emerging is thus one of an unenlightened, unruly agency steeped in rage and irrationality. This enduring trait of the anti-whaling discourse resurfaces in the recent "Mad Whalers Disease" caricature. On the other hand, reason, or rational morality, the quintessentially *human* attribute in the hegemonic Western moral discourse, is appropriated by "us." Second and closely linked is the opposition "barbarian versus civilized," another binary that ran deep roots in the history of East–West, and in particular Japanese–American relations (Campbell 1994). Examples abound in the contemporary media that feature the predication of "whaling" as "barbarian" as another lasting trope of the anti-whaling discourse. In 1992, a British Parliamentarian declared "if the Japanese like to eat 'exotic' food like whale meat, they should eat each other" (*Asahi Shinbun* 1992). The New Zealand Prime Minister commented on Japan's "barbaric" whaling in November 2000 (*The Christchurch Press* 2000). Interestingly, this dramatization of a devilish, bestial, agency is not without recalling the "intelligent malignity" Melville ascribed not to the whalers but to the *whale* in the archetypical American whaling epic.[7] Another cultural reference implicitly rewritten through the wielding of categories is the Christian cosmological order, specifically the hierarchy between angels, humans, and beasts. Whales, these "gentle and intelligent," indeed biblical creatures, exemplify here the best of what man can aspire to being, they are quasi-angels, and in lieu of the beast, now, is the "savage" (line 48) whaler.

These binary oppositions are the strategies of "othering" upon which the anti-whaling "self" is founded. Indeed, enhancing their "strangeness"/"foreignness" and the degree to which *we are not like them* is a powerful foil for asserting who we *are*. Their "difference" serves to reassert our identity. This was already a well-entrenched discursive strategy with regard to Japan: as Campbell (1994) has shown, much analysis of Japan's economic success was invariably predicated on Japan's being irreducibly *different*. Thus the boycott advertisement recalls the U.S. automotive industry advertisements' emphasis on the "differences" between the Japanese and Americans (such as height, for example; Campbell 1994, 147). One characterization becomes especially charged against a historical backdrop where Japan was the most important holder of U.S. debt: twice the lexicon of greed is used to describe the Japanese ("pure greed" line 67, and "the greedy Japanese whalers," line 80), whose dietary customs are explicitly evoked (lines 44–45). Hence the nascent anti-whaling discourse latched onto preexisting, effective strategies of "othering" deployed in other discourses of national identity. Furthermore, the attack on Japanese culinary practices has remained an enduring feature of the anti-whaling discourse. Twenty years later at the 2003 annual meeting in Berlin, IFAW displayed throughout the city an advertisement featuring a gigantic pair of chopsticks catching a whale as it sprang out of the ocean. The continuing resort to Japanese "strangeness," to their "cultural difference," tends to blur the line between "anti-whaling" and "anti-Japanese."

Against "Them": Constructing an Anti-Whaling "Us"

The text establishes a new set of relations between two types of subject-positions (Japanese and Soviet whalers; environmentally conscious American consumers) that rewrites the consumptive relation with whales—consumption, no longer to *kill* the whale, but to *save* it. These consumptive practices are further analyzed in the following chapter. In carving out the second subject-position, it defines a new type of universal, or indeed universalizable, moral agency, around individual consumptive practices. The text is designed to generate a wide constituency of consumer–activists. To this end two key strategies are deployed in tandem: strategies to enjoin and empower the consumer/reader individually and the creation of a collective "we" with which the reader/consumer is compelled to identify. I begin by considering the latter.

The creation of a unified "we the consumers" is a linchpin of the strategy. Thus eighteen of the largest national American NGOs stand here united in their call for action. This is both an exclusive "we" set up against a "them" and a "we" aspiring to humanity as a whole. This tension is wrought by two different strategies played out simultaneously. On the one hand, to create this new anti-whaling consumer's club, the text exploits a logic of belonging that functions as almost the exact inverse of Melville's "whaler club," which is the club of gods and heroes:

> Perseus, St George, Hercules, Jonah and Vishnoo! There's a member-roll for you! What club but the whaleman's can head off like that?" (Herman Melville, 1851 chapter 82, "The Honour and Glory of Whaling.")

In fact, the invocation of this classic of American literature within the advertisement (lines 84–87), which at the same time completely inverts the human–whale relation featured in the original text, reveals the anti-whaling movement as a mirror image of the nineteenth-century Yankee whaling culture, a sort of "Yankee particularism" turned on its head that took hold at a time a time of affirmation of "cultural particularisms" in the United States, in the late 1960s–1970s.[8] Moreover, this logic of belonging is not without recalling that of the "whalers' club" operating at the interstate level (see chapter 5). Central to this creation is distinguishing between an "inside" against an "outside." As Campbell (1994, 150) observes in his analysis of U.S.–Japan discourses, "the demarcation of an inside and an outside has an axiological dimension through which the social space of inside/outside helps constitute a moral space of superior/inferior."

On the other hand the "us" construed here lays claim to various attributes of universality. The text locates the "international community" (line 16) squarely within "our" camp, both as a moral community and a set of international institutions. Indeed, "the United Nations" (line 15), the "UN Conference on the Human Environment" (lines 15–16), and "the UN Food and Agriculture Organization" (lines 68–69) are all invoked here to predicate the world institutions as "anti-whaling." But this is also the "we" of a moral community, evoked by reference to a commonality of desires ("international pleas," line 18) and efforts ("international attempts," line 8). The text invokes a transnational, moral agency, a "global will," thwarted only by the Japanese and the USSR's evil deeds. This is a universality that, both in its humanist–moral and rational–scientific guises, tends toward the eradication of difference in the name of unity (Foucault 1988b, Balibar 1991, Walker 1993, Hardt and Negri 2001). Indeed, a sign of its incontestability, even the Soviet scientists are on "our" side (line 10). However, one remarkable feature of the anti-whaling discourse is that it combines morality, reason, *and* emotion, tapping into a very specific, Romanticist and Rousseauan strand of discourses of modernity. The anti-whaling discourse premises a universal human feeling for whales.

Polyphony, or the use of citation, is employed in the text to signify a rallying to the cause of a wide array of various voices of authority (see also Calsamiglia and Lopez-Ferrero 2003). First, scientific authority is deployed in several ways. It is personified twice, in the figure of "the specialist" Dr. Harry Lillie (lines 49–56), then, pulling in the mantle of fame, the oceanographer Jacques Cousteau (lines 88–94). Individual scientists add an emotional note to the scientific discourse. Dr. Lillie's "I" incarnates the rational witness before the horrors inflicted upon whales. Jacques Cousteau, another sea specialist, then *speaks for* the whale, who "could one day tell us something important, but it is unlikely that we will hear it" (line 93). Lastly, the unified voice of an institutional, universal scientific authority is evoked by quoting the report of the Scientific Community/FAO (lines 71–73). The text also taps into political authority. The figure of "Secretary of State Henry Kissinger" (line 40), whose reprimand to Japan (in a cable) is woven into the text (line 42), sets up a strong figure of leadership. Finally, literary authority is invoked by the figure of Herman Melville. The text (lines 84–87) singles out a quote that rewrites the classic as a tale of compassion for the whale, rather than a murderous confrontation between "man" and beast through which the American (man) made himself.

Thus a vast "us" is conjured up here against "them," encompassing the global civil society, the community of scientists, heads of states, literary figures, and the whales themselves—but stopping at the Japanese and Soviets. A dividing line runs through humanity, along a humane–human/inhuman–inhumane distinction. Most importantly, for purposes of interpellation and identification, the whales, those "gentle, intelligent" "agonizing" beings (line 29) are "like us." And thus summoned at the center of this vast "us" is the reader her- or himself, the ultimate subject of save-the-whale actions. The imperative form is deployed both to interpellate and to empower the reader: it is used already in the headlines, "save," "boycott," and the text closes on another set, "support," "don't buy," "tell your friends," "give generously" etc.; the text plays up the power of the American consumer. Another empowerment strategy consists in setting the example of a strong, authoritative agency, that of the U.S. government, who "strongly rebuffed" (line 81) Japan's suggestions. Even better, this strong agency is individualized, in the figure of Henry

Kissinger, who "sharply criticized" (line 40) the Japanese. Furthermore, the timing of publication compounds the sense of urgency. It coincided with the beginning of the Antarctic whaling season, emphasizing the need for action as the whalers go out "this week" (line 76). In constructing the need for adamant action, the text administers alternatively the threat of extinction (gloom) and the possibility of remedial action (hope). The reference to the California gray whales recovering from "the brink of extinction" (line 79–80) thanks to protective measures rings the hope note, for the key message is: what you do will make a difference. In addition, it is both highly rational *and* ... tax deductible (lines 100–101)!

The Anti-Whaling Amnesia: Transforming an *Antagonism* into an *Opposition*

This textual analysis offered a snapshot into a dominant discourse in the making. It considered how the anti-whaling discourse organizes agency so as to lock in a particular order of consumptive practices, which are further analyzed in the following chapter. The text revealed the anti-whaling discourse's processes of identity construction. The anti-whaling identity is founded upon a relation of opposition, where "the whalers-them" are set up as foil for the constitution of "our" selves, cast as ultimately ethical and reasonable global consumers. In the process it taps into strategies of othering already at work in other discourses on national identity, notably the U.S. political economic discourse on the "Japan problem." Another founding trope of the anti-whaling discourse that was brought to light was the monolithic construction of "the whale," which excludes other sets of signification deployed around the same animal in other cultures, not least Japanese (see chapter 11). This cleaving of agency between "them" and "us" also serves to rescind the notion of responsibility, between a historical responsibility on the one hand, and a moral responsibility anchored in the mode of action (present and future) but somehow shored off from the past. On the one hand, the construction of the anti-whaler rests on the active repression of Western involvement in whaling. On the other hand, it proposes a subjectivity that centrally relies on a moral agency—an actor who wants to "do the right thing." This construct thus shields "us" from any historical responsibility in order that "we" may be attributed the responsibility to act (now and in the future) to protect the whales from the whalers.

In order to construct such amnesia the anti-whaling discourse thus operates a maximum separation between "the whaler" and the "anti-whaler." It is driven by what Ernesto Laclau and Chantal Mouffe (1985) have characterized as a "logic of equivalence" rather than a "logic of difference," for the type of moral, globalizing agency offered by the anti-whaling subjectivity needed to gloss over many a local difference, in order to connect individuals from all the corners of the globe around the plight of the whale and against a common enemy, the whaler. Insofar as it was creating a new cosmopolitan, "humanist" identity that transcends local particularisms, it needed to establish a common point of reference that could draw together widely heterogeneous identities. The *logic of equivalence* establishes a chain of equivalence across logically different terms (or else there would be simple identity rather than equivalence). However, what holds the chain together is a common opposition to *what they are not*, to an Other. It works to simplify the political space (Laclau and Mouffe 1985, 130) and is thus well suited to the type of supranational political space that global environmental activism aims to carve out (see chapter 7). Since the opposition with the Other is essential to upholding the chain of equivalence that founds the Self, it tends to maximize the gulf between Self and Other.

What this maximal separation has achieved, however, is the transformation of an *antagonism* into an *opposition*. All identities are logically defined in opposition to what they are not: "to be A" necessarily implies "not to be B." In Baruch Spinoza's (1674) famous phrase, "every determination is a negation." His formula, however, captures a *logical* relation. *Political* relations on the other hand are rarely as clear-cut, and real oppositions rarely play out so neatly, precisely because identification is a dynamic process: political identities are neither fixed nor logically predetermined. They are always in the making, and their (re)production involves particular (re)configurations of power relations. This field of politics is thus marked by antagonisms rather than opposition (or, more specifically, rather than both "logical contradictions" of the type A–Not A, or "real oppositions" of the type A–B; Laclau and Mouffe 1985, 122). An *antagonistic relation* is one in which the presence of the other prevents the self from fully taking shape as itself. It prevents it from forming a closed entity, a totality (Laclau and Mouffe 1985, 125). Because, once again, the difference between "self" and "other" is not pregiven and is exactly what needs to be marked, what is at stake in the making of identifies, the other invariably threatens the constitution of the self. In the anti-whaling discourse, inasmuch as the other is actually *who the self used to be* historically, it is too close, disturbingly similar. The relationship is thus inherently antagonistic to start with. Yet what the anti-whaling discourse achieves is to constitute the other as *essential* to the making the self, but by opposition to it. The whaler is no longer an inconvenient reminder of who "the self" used to be, since all similarities have been performed away. The other has become the key term in the oppositional relation that founds the anti-whaling self. There could be no *anti*-whaler without a *whaler*. The anti-whaling identity was constituted by turning an antagonism into a logical opposition. It substituted a messy, antagonistic relation with a neat, binary opposition.

9

Crafting the Anti-Whaler (II): Consumptive Practices

The anti-whaler is defined not only by how she or he *speaks*, but also by how she *consumes*. As the boycott advertisement signaled, consumption went to the heart of anti-whaling activism. The end of whaling had entailed a complete reorganization of whale-related activities. Whereas manufacturers had simply turned away from whales once the petroleum industry could provide adequate substitutes for whale produce, anti-whaling activists sought instead to generate a new, *anti*-whaling economy, founded on a completely different relationship with the whales. Fostering a demand for whales *alive* rather than dead was, in their analysis, the best way to secure an interest in their long-term survival. This strategy was part of a broader and (since) rapidly developing trend in global environmental activism—one that has come under increasing critical scrutiny—to instrumentalize consumption as a key modality for greening human behavior (Durning 1992; Sachs 2000; Conca 2002; Princen, Maniantes, and Conca 2002; Epstein 2006). These anti-whaling consumptive practices are the object of this chapter. It analyzes the function of consumption in the structure of identity formation and belonging (the *subjectivity*) proposed to the individual in the anti-whaling discourse. It begins with an overview of the various "nonconsumptive utilizations of the whale" fostered by anti-whaling activism—ways of consuming the whales alive and whole, rather than sliced up.[1] This opening section offers a counterpart to the typology of consumptive uses that opened chapter 3. Because the passage from the whaling to an anti-whaling economy was inscribed within a broader shift from material to immaterial forms of consumption, the chapter then examines these consumptive practices against the broader evolutions of consumption patterns in globalizing postindustrial, information economies (Bell 1979, Castells 1996). Lastly, it sketches out the global anti-whaling consumer that emerged, almost a mirror opposite of the "whaling consumer" featured in the 1946 Whaling Convention.[2]

The salient theoretical feature that takes center stage here is the relationship between discursive and extradiscursive practices. Returning to points made in the opening chapter of this book, the definition of "discourse" adopted here is a "thick" one that encompasses not only things said but also things done, that is, practices, envisaged as meaningful actions. A discourse perspective focuses not primarily on words as such but rather on the production of *meaning*. Thus, it is concerned with any type of signifying practice, that is, *any practice that functions as a site for the production of meaning*. Insofar as the anti-whaler is shaped centrally by what she does and does not consume, consumptive practices have to be considered as well. This chapter is thus concerned with the structures of meaning—the semiotic structures—underlying these consumptive practices. The previous chapters already considered various dimensions of the production of meaning: the making of discursive objects (the "Superwhale"), of discursive subjects, and indeed identities (the anti-whaler). In this chapter, the way in which meaning emerges in the *exchange* between social actors comes under consideration. That is, exchange is envisaged a key site for the production, reproduction, and alteration of meaning. In economic interactions, more often then not what social actors are exchanging are not merely goods and services (that have an economic value) but also *signs* (that are invested with significations and social or moral values). It is the structure of this semiotic exchange that interests me here, and how it imbricates with the structure of economic exchanges in the anti-whaling economy. The anti-whaling economy is therefore both a "real" economy and an economy of signs.

Tell Me How You (Don't) Use Whales, and I Will Tell You Who You Are

Nonconsumptive uses are considered here according to whether the animal itself is involved directly (as a source of income) or only indirectly (as a symbol). A first set of whale-related economic activities has developed around interactions with "real" whales, mostly watching but also sometimes swimming with, either in the natural habitat or in marine parks. Marine parks represent the oldest nonconsumptive activity. They are historically significant as the site where the whale was made into an "entertainment" object. Here the wild animal is integrated into the urbanized landscape rather than enjoyed in its own environment. Increasingly controversial in conservationist circles, they nonetheless continue to thrive. Marine park whales all have familiar pet names, Jojo, Simo, Donald, Opo, Keiko the killer whale, star of the popular film *Free Willy*. An increasingly successful activity of late are the "dolphin assisted therapy" or "swim with" programs (either dolphins or small whales) proliferating in the United States and Latin America, proposed either with cetaceans in captivity or in the wild, also controversial among conservationists because of the negative effects on the animals.[3] Much less controversial are the whale-watching programs, which have provided key material leverage to the anti-whaling cause (Keck and Sikkink 1998, 23). One key NGO argument, that rapidly caught on with states as of the late 1990s, is that whale-watching had grown into a billion U.S. dollar industry—compared with the 60 million U.S. dollar value of the Japanese and Norwegian whaling industry (Hoyt 2001). It is also rapidly growing, with 4 million consumers in 1991 to over 9 million in 1998. There are more than 200 tours in North America alone. It is a high-level spending activity. In the early 1990s, a tour to Alaska with the Whale and Dolphin Conservation Society cost £3,000; to the Antarctic and Falkland Islands, around £5,500 for Alaska (Kalland 1994c). For a day trip in Andenes, Norway, the prices in 2004 were U.S. $150, and between U.S. $63 and U.S. $150 on the Californian coast.[4] As for nonconsumptive uses as a whole, 1980 marked the turning point, when their total value equaled the commercial whaling industry ($100 million; Kalland 1994c, 159). Thus by the early 1980, the discourse was so ingrained that it had already generated an alternative set of whale-related practices, ahead of their becoming entrenched into policy by the commercial whaling moratorium.

One type of "nonconsumptive utilization" hinges on both sets of use, that is, the "real" whale and the whale as symbol, namely, NGO recruitment practices. In the structure of NGO activism, the justification for member donations is grounded in their hands-on engagement to save "real" whales. This battle is always the ultimate reference point for all NGO activities, hence the importance of Greenpeace's anti-whaling maritime expeditions that continue to this day. However, in actual practice, the "real" whales are far removed from the urban settings where most of the recruitment takes place. There, recruitment functions on a closed circuit, where this external reference point fades out of sight: the symbolic value of the whale is what sustains or augments membership. And a potent value it is: Greenpeace, whose name was made on the whaling issue (Hunter 1980), counts 2.8 million members worldwide today (ten times the population of Iceland) and a total income of 158 million euros, an income comparable to that of a small multi-national corporation.[5] Greenpeace's success is a function of its ability to compel existing members to maintain their contributions and to convince new individuals to adhere. It has to prove that it "makes a difference," that being a member is worthwhile. Much like a large corporation with their shareholders, Greenpeace needs to provide reasons for continued stakeholdership. Like a business corporation, size is key to its success: "It was necessary for us to be big," commented the former head campaigner for Greenpeace Germany (Kalland 1994b, 7)—hence, in this logic, the importance of "winning" issues (see also Jordan and Maloney 1998). The "pragmatic philosophy" prophesied by the former head of Greenpeace International, Steve Sawyer, is one in which the battles chosen are bound to be won (Pearce 1991, 40). "We are strategic opportunists," declared Harold Zindler, head of Greenpeace Germany, to the journalists of *Der Spiegel* in the early 1990s (Schwarz 1991). To this effect it employs public opinion polling organizations to determine which issues attract the most support in which country (Schwarz 1991, 99). It uses the latest marketing techniques, carefully monitors the success of its advertising campaigns, and constantly adjusts them (Jameson and Eyerman 1989, 107; Jordan and Maloney 1998). In this vein the sealing issue, for example, has been dropped, in contrast with whaling, which has consistently been gauged a winning card. More broadly, Greenpeace's discursive strategy tends to deploy an upbeat tone and the rhetoric of victory in order to cultivate the self-image of a "winner."[6]

A second sphere of economic activities has developed around the commodification of the whale as icon. Here it is the image of the whale, rather than the real whale, that sells. It sells quite literally, in that whale photographs are a significant contribution to NGO revenues, specifically around the time of the IWC annual meetings.[7] The whale photographs are also at the heart of a new type of consumptive activity, the "adopt-a-whale" programs. For example, the Web site of the Whale and Dolphin Conservation Society invites the visitor to "adopt a whale," by which, for an automatically debited fee, she will be sent photos and regular reports on her whale. An important source of income, this scheme has recently been reoriented toward businesses instead of individuals, thus tapping into the new brand of "green marketing" that has increasingly accompanied the rise of environmentalism (Sachs 2000). NGOs have designed new fund-raising programs that cast the sponsoring of the anti-whaling movement as a "business opportunity" (a sign that "your business takes the environment and the 'green wave' seriously"). In the early 1990s the Norwegian oil company Statoil, for example, disbursed 1 million kroner to WWF–Denmark to use the WWF logo in their advertisement to increase their petrol sales (Kalland 1994b, 175). Besides photographs, "whale objects" sell, as illustrated by the proliferation of "cetacean artists" carving out all manner of effigies to this "whale cult" (Kalland 1994c). In addition, this whale iconography is indirectly exploited to sell anything from cuddly toys, key-rings, inflatable water toys, to t-shirts, stickers, buttons, bags, coasters—the list is endless. Lastly, the whale is exploited within the entertainment industry at large. Witness the proliferation films featuring whales (the 1986 *Star Trek IV: The Voyage Home*, *Free Willy* in 1993, and, more recently, *Whalerider* in 2003), CDs (the songs of the humpback have made regular appearance in record stores since 1970s), or computer games. The books are of two sorts: a number of "scientific" or educational books (*Dance to a Dolphin Song* by the "dolphin therapist" Horace Dobbs) or novels (whales become sci-fi characters in D. Brin's 1983 *Startide Rising* and L. R. Abbey's 1990 *The Last Whales*). The symbolic construction of the whale has given rise to a highly commodified culture, the *whale cult* (Kalland 1994c).

Consuming/Producing the Virtual Whale

The anti-whaling economy took shape against broader processes of economic globalization. Of course the whaling industry was also pitted at the global level. However, it also remained very much embedded in the national production structures and thus remained a national—even nationalistic—industry (see chapter 2). In the anti-whaling economy, by contrast, the space for the circulation of these anti-whaling goods and services—the consumptive space—was the globe as a whole. These anti-whaling consumptive practices were made possible by broader evolutions in patterns of economic consumption. They are inseparable from the exponential rise in the consumption of images, information, and cultural commodities, those "intangible goods" forming an increasing portion of the volume of traded goods in postindustrial societies (Bell 1979). They are intimately linked to new technologies of communication and information (Castells 1996). The growth of the Internet and of cable/satellite television has multiplied the number of outlets (or screens) on which images of whales (either "suffering" or "marvelous") can feature; more generally, it has provided the infrastructure for an ever more intense circulation of these digital commodities (see also Poster 1997, Escobar 1994, Tomlinson 1991, Garnham 1990). They coincide with a gradual shift from heavy industry to the service sector as the most important areas of production, both in terms of added value and employment. And within the service sector at large, they are associated with the explosion of particular clusters of activities, notably the entertainment, advertising, and communications industries expanding to a global scale, and with the boom of tourism, in turn, tied to the rise of accessible travel. All of these are brought into play in an activity such as whale-watching.

In contrast to consumptive uses, where whale *parts* were used, for which the whale needed to be sliced up, in the anti-whaling economy the whale is consumed as a whole, as it were, since it must be kept alive. In nonconsumptive uses, what is consumed is the *form* of the whale: the virtual whale. The whale is not consumed as a piece of meat or a raw material. Rather, drawing on the anthropological analysis of language, it is consumed *as a signifier*. For the whale has become a signifier for many things: communion with nature, a harmonious social organization, and other variations on the theme of "old values lost" (real or mythological). What counts now are not so much real whales or any particular whale (a humpback, a fin, etc.) but rather what "whales in general" have come to stand for within a particular system of values and meanings. Sure enough, the whale is out there, as an alibi, but far enough out at sea. There comes a point along this trajectory from signified to signifier where the signif*ied* fades out of sight; it is eclipsed as a referent and anchor to the signif*ier*. Indeed, *the* whale, as representation (whether as an object or as a digital image), is what circulates widely on the market and on the Internet (see also Epstein 2003). More broadly, this movement from the "real" whale to the whale as representation is the basis for the commodification of whales in the anti-whaling economy. Put differently, these whale objects (or images of whales) are not valued because of their concrete use (their "use value"), but because of what they represent. In the anti-whaling economy, the concrete "use value" of the exchanged commodities fades out of sight—consider, for example, the rather limited range of use of an inflatable rubber whale, compared with, say, the baleens in the whaling economy (see chapter 2).

Productive systems, for sociologists and anthropologists, are also signifying systems; structures of exchange and structures of meaning coincide (Escobar 1996, 2001; Poster 1997; Featherstone 1991; Jameson 1984, 1991). Thus the whale/signifier obtains within a web of signifiers pertaining to a particular society. It is these modes of exchange, and the broader culture–economy to which they pertain, that are under scrutiny here, to explicate the consumptive practices developing around the virtual whale. This, in turn, requires examining the role of consumptive practices at large in producing the "global, commodified subject" (Persaud 2001, 209). Or, to rehabilitate agency, they require examining the practices through which she or he produces herself or himself *as* a full-fledged subject in a late capitalist society. Consumption, which entails an exchange (the purchase of a commodity), is the defining mode of social interaction in capitalist societies. In the words of Daniel Bell (1979, 14):

> Capitalism is an economic–cultural system, organised economically around the institution of property and the production of commodities and based culturally in the fact that exchange relations, that of buying and selling, have permeated most of society.

Exchange value is the quintessential capitalist value, its "never ending augmentation," in Marx's (1990) own words, powering the deployment of the system, and consumption is a key collective practice of capitalist societies (see also Jameson 1984, 1991; Poster 1997).

At this point, however, it is Jean Baudrillard's analogy between consumption and language that becomes especially useful for accounting for the consumptive practices at play in the anti-whaling economy. For only when we envisage consumption as a language can we explain how an object of little utility sells. Baudrillard develops a theory of economic consumption by drawing together a Marxian analysis and the Saussurean theory of language. For language is a structure of differentiation. That is, it is "not an absolute, autonomous system, but a structure of exchange contemporaneous of meaning itself" (Baudrillard 1972, 74).[8] Underpinning the analogy between these two types of structure is an equivalence between "meaning" and "exchange value," which hinges, in turn, on ambivalence inherent to the term "value" (an economic or a moral value). The progressive eclipse of use value in favor of exchange value is not unlike the loss of the signified in Saussure's thinking. For an object's use is its concrete referent; the disappearance of use value entails that the value of a commodity becomes determined exclusively in the exchange. Exchange relations are the locus of this production of meaning/value. At the same time, the purchased commodity serves to "mark" the purchasing subject—as a parent, a student, a sportsman, a reader, etc.—and establishes her position within the system. It defines the consuming subject, in terms of what she or he likes, how she spends leisure time: in short, in terms of who she "is." The commodity is a marker of her subjectivity and "individuality," that is to say, her "difference." In consuming, the subject "speaks" herself *as* a specific subject. She produces herself meaningfully in a system of exchange relations, that is, both as an individual and as a member of *this* society. As Randolph Persaud (2001, 209) puts it, the circulation of commodities operates as a signifying chain that brings the self and commodities into a social relationship. The symbolic order into which the subject is buying is also a social order. Thus consumption also fulfills a fundamental regulative function, insofar as, for Baudrillard (1972, 74), the production of difference (of "distinctive material") is a key modality of social regulation. Every purchase of a virtual whale (under some form or another) inscribes the consumer within the larger social–symbolic order, with its cluster of meanings and values written onto the whale. In buying her cuddly whale she subscribes to it and further perpetuates it.[9] By the same token, she establishes herself as a member of this society, a late capitalist, antiwhaling, globalizing society which values whales and devalues whalers.

Consumption is the "productive force" of the system (Baudrillard 1970). In fact, production itself has given way to *reproduction*, and images, signs, and information are the new "raw materials." This is the era of *virtualization*, or, in Jean Baudrillard's term, *simulation*. A "system of objects" has increasingly come to be replaced by a "system of signs," as the products are increasingly removed from any concrete finality (Baudrillard 1975, 87–93). Here too an economy of whale-derived objects that pervaded everyday life has been replaced by an economy developed around an omnipresent whale *sign*. At this point, a new distinction is needed to elicit yet another dislocation of the signifying chain: no longer a *signifier* but a stand-alone *sign*, given that the signifier, increasingly disconnected from the "real" signified, has come to refer to nothing but itself. Whereas the signifier continued to maintain some relationship with concrete whales, the sign no longer does; the relationship is purely arbitrary. The epicenter of production and exchange has shifted from material objects to signs and their endless repetition (see also Jameson 1991). At this point what Baudrillard captured as "logic of the code" takes over, with its empty, self-referring dynamic. The code operates on a model of simulation: the real is cut out to the benefit of its "simulacra." Likewise, the anti-whaling economy thrives on the consumption of whale simulacra. This centrality of simulacra to this economy of signs is aptly rendered by the whale probe featured in the 1986 film *Star Trek IV*: in the twenty-third century, planet earth is threatened by a gigantic, alien probe shaped like a whale and emitting whale sounds. The earth is saved by humpback whales, whom the Enterprise crew set out to retrieve from the past and who are able to communicate with the probe for it to leave the earth.

Simulation, for Baudrillard (1988a), functions not merely on the disjuncture from the real but on *its absence*. Simulation stipulates the disappearance of that which is simulated, all the while masking this absence. Nature, the real whale, needs to be held at bay in order to be recreated as a virtual whale. In other words, this distance, which, for Baudrillard, is operated by the sign, is part and parcel of the anti-whaling economy. Extending this line of thought, the question arises of to what extent the "virtual whale" feeds on the disappearance—real or construed—of the real whale, that is, to what extent does the commodification of the virtual whale require positing whales as "endangered." Indeed, this postulated endangeredness is a cornerstone of a whole series of marketing practices that have spun out of the "endangered species" signifier (see chapter 5). It is key, for example, to the selling of the Endangered Species Chocolate Company's chocolate bar across supermarkets on the United States' West Coast, or to the Endangered Species Shop on San Francisco's Embarcadero, which sells anything from cuddly whales to pink gorillas to the visitors of the city's tourist district. These practices fall into what Frederic Jameson (1991, ix) would evoke as a postmodern culture–economy, or "what you have when the modernization process is complete and nature is gone for good."

This heralds the hegemony of "the code," that is, for Baudrillard, beyond the discourse of *choice*, "a structure of power and control" centered on the individual, with consumption as its key regulatory mechanism (Baudrillard 1975, 87). Again, this is social control at its most intimate, operating at the level of the individual, the "atom of society," (Foucault 1981, 1010). The individual is identified within the system as a *consumer*, in other words, a combination of needs, codified according to the different role she or he plays in the system: as worker, parent, student, leisure-time consumer, etc. Through the spread of the consumptive models, human needs are progressively standardized within one code, rid of the arbitrariness of desire, and thereby effectively socialized. Overemphasizing the argument slightly and in Baudrillard's (1970, 42) own words, "Needs are produced as a force of consumption, and as a general potential reserve, within the larger framework of productive forces.[10] Consumption thus becomes an area of vital strategic control—hence the explosion of advertisement (Baudrillard 1988a, 1988b). Capturing demand, preempting it, articulating it, sustaining it, effectively directing it, becomes the essential motor of a globalizing production. Within a productive logic where ever more consumers need to be generated in order to sustain the expansion of the system, the monopoly of consumption functions as a system for generating, and subsequently managing, demand. And insofar as these dynamics capture the regulation of a *global* order, they are intimately entwined with interstate relations, as we shall see in the following chapter. As Randolph Persaud (2001, 209) underlines, "at the level of international relations, transnationalization *per force* involves the *homogenization* of commodified desire on a global scale." The making of a global commodified subject thus shapes foreign policy and interstate relations as well. If what the consumer wants to buy is a picture of a whale rather than a whale steak, then the state to which she belongs must be anti-whaling and oppose whaling in other countries too.

Consumption is a system for simulating *and* stimulating demand. In this perspective, then, the so-called green demand that upholds the anti-whaling economy as well as other environmental activist practices appears altogether delinked from any "essential" human need. Instead it is generated, sustained, and managed within a code. NGO membership recruitment practices appear, in turn, as practices of socialization and normalization. First, they operate as *individualizing* practices: the adopt-a-whale scheme, for example, matches one person to a whale, christened and personified in the "adoption reports." The target for these personalizing strategies is the individual, whose "individuality" is reinforced by this "personal relationship" with a whale. They exemplify Foucauldian technologies of individualization that serve to position the good, moral, and ultimately "normal" individuals (Foucault 1975, 1990, 2003). These fund-raising practices contribute to constructing what Persaud (2001, 209) has dubbed "marketed common sense on a global scale." The composition of NGO membership is revealing in this regard. In 2002, 93 percent of Greenpeace's income came from individual donations (Greenpeace 2004b).[11] While tailored for individuals, the promise is of membership of a global moral community, as we saw in the previous chapter, for Greenpeace has successfully generated a "global village" or virtual community, organized around the Internet, and thriving on a new electronic intimacy binding together individuals scattered around the world, an apt illustration of Manuel Castells's (1996) "network society." It invokes what sociologists have characterized as a form of late-modern belonging, a belonging that is no longer anchored to a common place and common history but operates instead on a series of "distanciations" and ruptures with these traditional attributes of belonging (Giddens 1990). This "reflexive modernity," as Ulrich Beck (1992; Beck, Giddens, and Lash 1994) has described it, is one where the notion of individual choice becomes acutely foregrounded. That is, in a setting where the expectations of traditional institutions hold less and less sway, increasing weight is brought to bear on the individual's personal choices in shaping the self. Hence late modernity is also a place of "risks," associated with specific forms of "individualization," notably in politics. Likewise, the consumptive practices described here point to a new type of "communities of individuals" from all over the globe, a manner of anti-whaling e-community founded in distance and mediated communications rather than proximity and immediate interactions.

The Global Anti-Whaler

Who, then, is the subject conjured up within this moral–economic system? The *global anti-whaler* is the typified target of anti-whaling marketing.[12] She or he is an urban dweller in a developed country. Theirs is an increasingly *mediated* experience of the wilderness that contrasts with the immediate, confrontational experience of nature invoked in whaling (see also Epstein 2003). They would tend to be Internet users; at any rate, they would recognize themselves as a "global citizens," members of the "global civil society." Their leisure activities (hobbies) would involve some form of whale-related consumption (watching a movie with whales, purchasing the "songs of the humpback whale," etc.). Here again Baudrillard's perspective becomes relevant. The "sophistry of consumption" is such that leisure itself is central to the mechanics of the system: not only is leisure the "time" for consumption but leisure *is* the consumption of unproductive time. Far from being passive, this is activity, an obligatory social ritual. Thus being a "[wo]man of leisure" is part of being integrated to the modern urban society. As Baudrillard (1972, 760) writes:

> Time is not in this instance "free," it is sacrificed, wasted: it is the moment of a production of value, of an individuous production of status, and the social individual is not free to escape it. No one needs leisure, but everyone is called upon to provide evidence of his availability for unproductive labour.

The anti-whaler is also a traveler, someone who has seen the world. Traveling is also a modality of leisure consumption. For Dean MacCannell (1999)—who, in a similar vein, deployed an "ethnography of modern society" through the prism of the phenomenon of tourism—along with work leisure has become central to modern social arrangements: "The empirical and ideological expansion of modern society [is] intimately linked in diverse ways to modern mass leisure, especially to tourism and sightseeing" (MacCannell 1999, 3). The explosion of tourism ties in with the emergence of a new "international middle class," the "leisure class." Thus, for MacCannell, tourism is the quintessential modern experience, sight-seeing the ultimate modern ritual (MacCannell 1999, 42). Tourism implicates a moral order: it requires a "moral engagement" on behalf of the tourist to whom the spectacle is offered (for example, slavery monuments are expected to prompt indignation, etc.). This moral dimension is also present in whale-watching off the coast of the Dominican Republic.

An additional element to the tourist experience is the proximity with extinction. For MacCannell (1999), tourism is the ultimate consumptive experience, grounded in the consumption of signs—typically, monuments such as the Eiffel Tower, the Empire State Building, are consumed, not literally of course but as signs. Once again, these signs feed on the eclipse of the signified. Anthropologist Marie Françoise Lenfant for her part has revealed an inverse correlation between the touristic value (of an animal, a piece of folklore, etc.) and to how close it is to becoming a petrified form, a relic of the past. In other words, what makes the attractiveness of a touristic object is its proximity with death (Lanfant and Graeburn 1994). It is because whales are on the verge of extinction that whale safari tours are rated so highly, according to a marketing logic grounded in the risk of extinction.

This chapter began with a typology of nonconsumptive uses of the whales that mirrored the survey of consumptive uses with which this book first opened. These ways of not consuming the whale were then analyzed against a broader evolution toward image-based, immaterial forms of consumption revolving around virtual rather than "real" commodities, in our postindustrial, service-based, virtual economies. In the anti-whaling economy, what is consumed is a virtual whale that has to be kept whole, as opposed to the parts of a real, sliced up whale. In other words, what is consumed is a whale *sign*, a set of meanings and symbols that befall the consumer in her consumptive act. In other words, in purchasing this whale-related virtual commodity, the individual consumer is marking herself as someone who cares about the environment. This, in turn, led me to analyze the social function of consumption as a system for generating, managing, and ultimately normalizing demand. The chapter ended with a rough sketch of the highly individualized ideal-type identity thus generated within this consumptive system: the anti-whaler is a consummate Internet browser, dedicated to traveling, who cares about the environment.

10

State Positionings (I): The Anti-Whaling Discourse

Whereas the focus in the last two chapters was cast on the individual, this and the following one are set at the other end of the levels-of-analysis spectrum, upon the state. As we saw in those chapters, the anti-whaling discourse primarily interpellates the individual who, in stepping into its subject-position, is marking herself or himself as a good, cosmopolitan citizen who cares about whales and the environment by extension. But what exactly is in it for states? Why do *they* continue to care so much about whales? State attendance at the IWC annual meetings has been consistently on the rise, indicating an unwavering interest on behalf of states. Membership has jumped from twelve to seventy-seven states by 2007, increasing by around five or six annually (five new members in both 2006 and 2007, six in 2005, seven in 2004, three in 2003). Yet the IWC is an uncooperative (Friedheim 2001a, 2001b), even a rancorous place. Since the ten-year period for which the commercial whaling moratorium was originally scheduled came to an end in 1992, debates have become increasingly polarized between a loud majority of anti-whaling states and an increasingly louder minority of pro-whaling states. Nor is much achieved in policy-making terms. Because any significant policy change—such as the overturning of the moratorium, or the adoption of a new whale sanctuary in international waters—requires a three-quarter majority vote, and since both sides are now large enough to block the formation of such a majority, policy making has all but ground to a halt. Each year, the IWC meets to uphold an expired moratorium that bans the activity it was originally set up to regulate, thereby further locking in the anti-whaling order. And invariably each year the question is, how much ground has the pro-whaling minority recovered away from the anti-whaling majority? The IWC is thus largely an organization looking for a mandate. There have been attempts to shift the policy focus to related, secondary topics that do not require three-quarter majority votes, such as whale-watching, but many of these come up against the limitations imposed by national jurisdictions. In sum, this unwavering state interest in the IWC is not about successful cooperation, nor is it about productive policy making. Rather, it is about positioning oneself within a society of anti-whaling states. It is a social act of self-definition.

State Subject-Positions and State Identities

This chapter and the next examine the play of state positionings at the IWC, by which I mean the ways in which states draw on discourses on whales and whaling to position themselves on the issue. This chapter considers the anti-whaling states; the next turns to the pro-whaling states. By taking up the anti-whaling discourse, a state is marking itself as an anti-whaling state, both before other states at the IWC and, given the level of media attention the issue attracts, before the world beyond. Hence the concept of identity is implied in this focus on state positionings; however, it is so only in a very narrow sense that requires clarification. First, it is a "thin" rather than a "thick" identity rooted in historical structures that have shaped the nation and national interest. Indeed, for some of the loudest anti-whaling states, many of which were formerly whaling states, their prior involvement in whaling had a more significant impact in shaping their national interest than their being anti-whaling (see chapters 2 to 4). The distinction I am drawing here, without wanting to dwell on it too heavily, is between *state* and *national* identity, the latter only really applying to pro-whaling states as we shall see in the following chapter. A key reason for this has to do with the anti-whaling identity itself, which is foremost an individual identity (see chapters 8 and 9), projected, furthermore, *beyond* the nation-state, as a cosmopolitan deterritorialized identity (see chapter 7). Hence it makes little sense to evoke British or Australian varieties of anti-whaling-ness. Nor as a result are these anti-whaling states studied individually, unlike the pro-whaling states in the following chapter.

The other main reason has to do with the concept of identity itself. The anti-whaling identity lends itself especially well to revealing the open, fluid, flexible, even contradictory nature of identity. That the very same state, such as the United Kingdom or Australia, can suddenly switch from being a whaling to being an anti-whaling state illustrates the fundamental *unfixity* of identity. It moves identity decisively away from any essentialist underpinnings (since being anti-whaling is delinked from inherent attributes of British or Australian-ness) and away from historical determinism. It shows not just that identity is discursively construed but that it is a performative and hence willful act of identification. This is key to understanding how, despite a whaling history, a state such as Australia could swiftly embrace a new identity largely imported from without (see chapter 7). A state, just like an individual (see chapter 8), can take up different, even opposite identities. This does not mean that being anti-whaling is an insignificant identity for Australia. Indeed, judging by the virulence of Australia's anti-whaling statements (Dorney 2005), it would certainly not seem to be the case, even if it is perhaps not its most significant identity. Nor does it mean that Australia is somehow feigning an identity it does not really believe in. *A state is what it says it is*, since, in speaking, it is performing itself. The issue, in terms of analyzing the construction of state identities, is thus not to what extent the state actually "believes" in its anti-whaling statements, or whether it is just "pretending" in order to achieve something else (even if this may well be the case), but simply the fact that it is proffering them. This performativity of identity elucidates this apparent contradiction by which a state such as the United States can still host whaling upon its territory yet discursively position itself as an anti-whaling state. It points in addition to the other crucial feature that is especially emphasized by the anti-whaling identity, namely, its *relational* dimension, for being seen as being on the "right" side by social others within a society of states largely regulated by an anti-whaling norm is a key driver in states' speaking the anti-whaling discourse as we shall see. These social dynamics at play in the construction of identity confirm the need for a sociological approach to interstate interactions for which the analysis in chapter 4 paved the way.

Having unraveled the relationship between state positionings and identities, to avoid any confusion that may remain in wielding the concept, I now suggest letting identity fade into the background altogether and setting the lens instead upon *subject-positions*. The concept of subject-position, ushered in by focusing on discourse, circumvents many of the difficulties that stem from transferring the concept of identity to the international level. It offers a way of studying state identities that excludes the dimension of *subjectivity*, which is a notion that only applies properly at the level of the individual (see chapter 8). For subjectivity, with the figure of the modern political subject, lies at the heart of conceptualizations of identity in Western political thought. Thus it carries with it certain characteristics—specifically, bodies and affects (Butler 1997)—that are not so readily transferable to collective political units, let alone states. Shifting the focus to subject-positions instead of identity avoids the awkwardness that may arise when identity is wielded as the central concept to bridge the levels-of-analysis problem in international relations (Wendt 1999). From there indeed it follows that one has to assert that, in Wendt's own words, "states are persons too." This, in turn, opens up a whole set of conundrums such as how much are they *like* people? Do they have feelings too? Indeed, when do we know when they are genuine? Should we then call upon psychologists to study them? (See, for example, Mercer 2005.) States need not be "persons" to have an identity, nor do we need the tools of psychology to analyze them, *so long as* identity is defined away from subjectivity.[1] The discursive approach offers a way of salvaging constructivists' central concern with identity by orchestrating two key moves. First, and to repeat an important point, it draws out identity as a dynamic, perfomative process of *identification* rather than identity (Zehfuss 2001, Hansen 2006). Second—the move operated here—it shifts the focus to subject-positions. When it comes to states, subject-positions constitute identities *minus* subjectivities.

Since the IWC is also a policy-making body, how do these identity-related concepts relate to the concepts that help us understand policy making? These state subject-positions are also policy positions, or rather the former are the basis for the latter.The specific policy prescriptions advocated by a state at the international level stem from broader positionings on the issue, determined, in turn, by stepping into a particular subject-position. Hence subject-positions both found, and provide a principle of coherence for, the states' policies. Consequently, as in chapter 5, the tools developed by the discursive policy-making literature continue to remain useful for unpacking both the anti- and pro-whaling state discourses: namely, *story-lines*, or the narratives that are drawn to constitute particular policy positions and subsequently to justify them, and *discourse coalitions*, which captures the regrouping of disparate actors around a common discourse (see chapter 5). However, the policy focus is not the most pertinent angle of analysis with regard to the IWC, not least because of its current policy limbo. The IWC today resembles more a vast arena for the confrontation of two discourses than a productive policy-making forum.

The "Antagonism of Strategies"

States carve out their subject-positions by deploying particular discursive strategies. Insofar as they are key to state positionings, these discursive strategies are the main focus of these two chapters. They are structured around the central confrontation shaping the debates at the IWC, between a dominant anti-whaling discourse that works to perpetuate its hegemony (analyzed in this chapter) and a dominated pro-whaling discourse of resistance that seeks to dethrone it (analyzed in the next). Hence the antagonism of these two sets of strategies is the organizing principle of this analysis of state positionings. Methodologically, this builds upon Michel Foucault's injunction to target struggles as the privileged site for an applied analysis of discursive power:

> I would like to suggest another way to go further toward a new economy of power relations, a way which is more empirical, more directly related to our present situation, and which implies more relations between theory and practice. It consists of taking the forms of resistance against different forms of power as a starting point. To use another metaphor, it consists of using this resistance as a chemical catalyst so as to bring to light power relations, locate their position, and find out their point of application and the methods used. Rather than analyzing power from the point of view of its internal rationality, it consists of analyzing power relations through the antagonism of strategies (Foucault 1982, 208).

The points of friction between the two discourses are thus key to drawing out how discourses are fundamentally imbricated with power. The clashing of quills, or rather words, on the floor of the IWC serves to reveal the power relations between these discourses. One of the key advantages of the whaling debate is that, because the dominated discourse is relatively assertive, the lines of battle are drawn out in the open. The pro-whaling discourse thus serves as a catalyst for exposing the operation of discursive hegemony. What renders a discourse "dominant" is that it has successfully evacuated alternative frames in imposing its own. Dominated discourses remain the repositories of these alternative sets of meanings; however, they are by definition not readily accessible, unless they have begun to mobilize. In some ways the very polarization of the IWC, although it serves to entrench the status quo and therefore perpetuate the anti-whaling discursive hegemony, is also an indication of some measure of success for the resistance of the pro-whaling discourse, at least in terms of stalling the progression of the anti-whaling order (for example, by successfully defeating the adoption of new whale sanctuaries).

This chapter maps the various strands of the anti-whaling discourse and how it has changed over time, in reaction to developments at the

IWC, to the consolidation of a pro-whaling discourse, or by connecting to broader metanarrative; insofar as considering discursive strategies also requires examining how discourses connect to other discourses. As we saw in chapter 5, discourses can become powerful by tapping into well-entrenched, taken-for-granted metanarratives—by drawing on the power of common sense. The chapter observes how the anti-whaling discourse served to carve out subject-positions for states as "good" members of the society of states, around two broad sets of articulations. First, being anti-whaling serves to mark a state as green, ethical, democratic, *and* civilized. Second, being anti-whaling and being a performant, neoliberal state at the cutting edge of globalization increasingly coincide.

Constructing the Democratic, Ethical Green State

Whales Are Endangered

Throughout the late 1970s and early 1980s, states' stance against whaling was straightforwardly cast as a conservationist argument about the need to protect endangered species. As we saw in chapter 5, the 1970s had witnessed the rise of endangered species protection as a global metanarrative and associated regimes of international cooperation around, notably, the 1971 RAMSAR convention on Wetlands of International Importance, CITES in 1975, and the 1979 Bonn Convention on Migratory Species (see also Epstein 2006). In signing up to CITES, states had signed up to the idea of safeguarding the species that were deemed endangered. Thus in a context where whales appeared to be disappearing to the naked eye (see chapter 2), endangered species protection provided a common ground of concern for many states, including states with no prior involvement in whaling but that were committed to protecting endangered species, such as Sweden. This explains the widespread state support for the commercial whaling moratorium in 1982.

The ten-year suspension of commercial whaling had a twofold purpose: to allow the recovery of overexploited whale stocks and to enable the scientists to ascertain more precisely the levels of endangeredness of the various stocks and, from there, to establish a new procedure for managing whale catches. By 1992, these efforts were well under way, such that the SC was able to present its first estimates of particular whale stocks that year together with their new management procedure, the RMP (see chapter 6). These results appeared to indicate that certain stocks of whales, such as the minke whales, which numbered 761,000 in the Southern Hemisphere alone, could not qualify as "endangered." At this point, whaling countries were in a position to argue that the ten-year whaling moratorium, whose term was also coming to expiry, had served its original purpose, at least with regard to some species, and that some whale stocks were healthy enough to be able to sustain controlled harvesting. The important point here is that, from then on, the whaling states could rest anew their case in favor of whaling without having to challenge the broader metanarrative within which states had framed their anti-whaling stance. They could take them up on their own grounds.

The publication of the 1992 SC report thus marked a turning point in the respective positionings of anti-whaling and whaling states. Notwithstanding its best efforts to steer clear of IWC deliberations and uphold its neutrality, the SC's work had decisively reshaped the political debate. Indeed, when the scientists presented their report on the floor of the 1992 plenary meeting, they refrained from making any pronouncements as to which species or stocks were endangered. Instead they limited themselves to merely announcing the results for each stock of whales for which calculations had been completed. Yet in doing so, they had also thrown back the discussion of "endangeredness" into the political arena, effectively confirming "endangered species protection" as the discursive terrain where whaling and anti-whaling states would confront each other. However the abundance of certain whale stocks revealed by the scientists' results made it difficult to uphold that *all* whales were endangered, hence that a blanket moratorium was needed. Thereafter the IWC was seen by some countries, in the words of the Guinean Commissioner, as being in the business of "protecting species not endangered" (IWC53/10 2001). The minke estimates, for one, were fodder for Norway's proposal to down-list the species from the CITES Appendix I's "list of most endangered species," where it had been listed since 1983 to allow some measure of international trade. Although it was systematically defeated in 1994, 1997, and 2000, the whaling counterattack had been mounted, using the very terms of reference of the endangered species protection discourse. An unintended political effect of the scientists' report, anti-whaling countries had begun to lose ground to the whaling countries on this discursive terrain.

At this point, the hitherto unified state anti-whaling discourse branched out in two different directions. On the one hand, as we shall see below, it started developing new story-lines. However, because there still remained significant symbolic benefits to be reaped from tapping into a dominant discourse, it also worked to uphold the "endangered species" story-line. In step with NGOs, anti-whaling states deliberately rejected any attempt to draw distinctions between the various stocks of whales during the discussions at the IWC. For example, when Australia and New Zealand proposed that the South Pacific be proclaimed a whale sanctuary at the 2002 IWC Plenary meeting, they were careful to list all the species and to specify: "There is no scientific justification for differentiating between or excluding any of these species ..." (IWC/54/16 2002, 3).

What is interesting is the way in which science is marshaled here: the proposal is *not* suggesting that there is no scientific justification for distinguishing between endangered and nonendangered stocks. Rather, displacing the resort to scientific justification, it is saying is that, *in the context* of discussing whale sanctuaries, there are no scientific reasons to exclude any species. In this way the loaded signifier of "endangered species" can be maintained as the linchpin of these states' anti-whaling positioning. This position relies on the taken-for-granted status of the category of "endangered species"—no one questions whether endangered species should be protected. Thus the broader strategy at work here consists in forestalling any recognition that not all whales are endangered. Discursively, this requires preventing any category that would expresses this recognition from taking shape, such as an opposite category of "non-" or "de-endangered." Thus the anti-whaling strategy relies on the discursive absoluteness of "endangered species" as a semantic category—the fact that it is not brought into a syntactic relation with a semantic opposite, or antinomy. Indeed, one seldom hears, or expects to hear, talk of "healthy" or "nonendangered" whales—somehow that expression does not ring quite right.

Whales Have a Moral Right to Life

Progressively, however, once "endangered species protection" had become a less certain terrain, anti-whaling states shifted to different grounds altogether. Thus the conservationist argument about protecting endangered species increasingly gave way to an ethical argument about whales' generic right to life. By the early 1990s, the anti-whaling discourse was so well entrenched that it successfully carried over an argument that had originated in the remote and habitually marginalized confines of animal welfare activism into the mainstream of international law (Scarff 1977, M'Gonigle 1980, Birnie 1985, Falk 1989, Gillespie 2001). For example, in 1991 the highly respected *American Journal of International Law* published a since much quoted article entitled "Whales: Their Emerging Right to Life," jointly written by a staff attorney for the U.S. Environmental Protection Agency (Sudhir Chopra) and an eminent Professor of International Law and member of their editorial board (Professor Anthony D'Amato), which articulates whales' entitlement to life as established *opinio juris* of customary international law. They write, "The entitlements of whales to live and be left alone has arguably resulted from the developing practices of various institutions—international, conventional and national—concerned with whaling" (D'Amato and Chopra 1991, 28). That prominent international lawyers should hold up whales' "right to life" as the beacon of a "broadening of international consciousness" that is fundamentally transforming international law (D'Amato and Chopra 1991, 22) created, in turn, new opportunities for states to polish their moral credentials by establishing themselves as guarantors of this new form of posthumane, or rather *humanist*, international legal right, to use their own term (D'Amato and Chopra 1991, 61, emphasis added). Almost two decades after this proclamation, no other species has, to my knowledge, been able to benefit from this proclaimed expansion of the "right to life" under international law—in fact international lawyers are often still struggling to have human beings recognized as subjects of international law.

The conservationist argument about overexploited species thus gave way to an animal welfare argument that posited whales as the new recipients of a universal "right to life," and, from there, posited states as the guardians of this new transspecies right. One such state was the United Kingdom, which, once the scientists had presented their estimates in 1992, shifted from a position of uncertain waiting to a principled opposition to whaling in general, steeped in arguments about animal welfare. The concern for animal welfare continues to characterize most United Kingdom interventions on the floor of the IWC to this day. For example, it has consistently championed the issue of humane whale killing, even while the "humaneness" of other killing practices such as fox hunting were still being debated at home.

Saving Whales Is Civilized

This principled posture in defense of whales readily slips into the language of Barbary versus Civilization, echoing once again the NGO discourse (see chapter 8). In this particular twist of the state anti-whaling discourse, the end of whaling in the West becomes equated with civilizational progress: for example, Jack Metcalf, congressman for the State of Washington, home to the Makah whalers, declared to Congress in March 1998 that "protecting whales has become one of our civilization's most noble undertakings." His opposition to the Makah's request of a gray whales quota was justified precisely on these grounds. On the other side of the Atlantic, although it was never explicitly formulated as an official condition of accession to the European Community, "nonwhaling" rapidly became an implicit marker of whether states were suitable to join the "civilized" club of Western European nations thus taking shape. Whaling invariably became a heated topic of debate both in the European parliament and in national parliaments when Norway or Iceland's accession to the EEC came under discussion.[2] Thus when Norway's membership was back on the table in the early 1990s British Parliamentarian debates, the agricultural minister, John Gummer, was hailed by fellow MP Simon Hughes in the following terms: "Does the Minister further agree that ... no argument that allows commercial whaling to continue is acceptable to this community or to any other country that calls itself civilised?" (House of Commons 1991). Indeed, he did. Anti-whaling states thus claim for themselves the mantle of civilization and relegated whaling states squarely into the category of barbary.

Saving Whales Is What the People Want

Another important aspect of the construction of the ethical state is the portrayal of oneself as democratic. Thus a central claim for anti-whaling states is that they represent the public sentiment regarding whales. Prior to the 1993 IWC annual meeting, where the resumption of commercial whaling was likely to be discussed for the first time, the official U.S. stance was wired out by the State Department to other IWC member states: "Many people in the U.S., including virtually all animal protection groups and some large environmental groups, oppose the resumption of commercial whaling" (quoted in Christoffersen 1994). Anti-whaling states are thus truly democratic because they speak the voice of their citizens, which is by the same token performed as homogeneous. Here the state is cast as the conduit for carrying through the concerns of the individual citizen to international forums. To this effect the alliance between anti-whaling states and NGOs is key. The anti-whaling states' association with voluntary organizations of citizens concerned with saving the environment serves to buttress not merely their green but their *democratic* credentials. After all, the denunciation of a democratic deficit is inherently tied up with the rise of popular environmentalism across the developed world in the 1960s (McCormick 1989, Pearce 1991, Guha 2000). Saving the whales is thus a way of both addressing this old environmentalist claim and catering to what the people at large want.

Furthermore, because these NGOs have successfully deployed themselves on a global level, the conjoining of states and NGOs serves to blur the lines between national publics (see chapter 5). Hence it casts anti-whaling states as listening not only to their own publics (the British, the Australians, or French) but to a "world opinion," construed as a global cosmopolitan citizenry of anti-whalers (see chapters 8 and 9). With the anti-whaling campaign, environmental NGOs have successfully pitched themselves as the direct representatives of new "global and universal human interests" that are seen as overriding the more traditional, indirect representation by states (Hardt and Negri 2001, 313). Armed with a global individual membership that extends even into some whaling countries, NGOs such as Greenpeace or the IFAW can claim to voice the popular sentiment, the "generic feeling" for whales.[3] Thus, in listening to NGOs, states portray themselves as in step with a "world public opinion."

One instance where this global state–NGO alliance was used to signify the state's ability to listen to "the people" is the Southern Ocean Whale Sanctuary, first presented by France to the IWC in 1992 (adopted in 1994). France then stood in rather unfavorable light with green NGOs, and indeed the international community at large, in the fallout from the *Rainbow Warrior* scandal (the blowing up of the Greenpeace ship by French security forces in New Zealand waters in 1985). By 1991 New Zealand, in the face of French recalcitrance, had dropped its extradition request. However, diplomatic relations between the two countries remained rather strained, and France's international green credentials undermined. The 1992 IWC proposal on behalf of a country that, while a founding IWC member, had characteristically remained disengaged was thus an inexpensive way of repairing its relations with New Zealand by doing something about an issue that mattered to the latter. It was also an attempt to buy back Greenpeace's support and the green vote more broadly at a time of its increasing political salience in the run-up to the second and largest UN environment conference (the 1992 United

Nations Conference on Environment and Development; UNCED). Indeed, the document that France had tabled at the IWC was a photocopy of a fax sent by Greenpeace International—it still bore the organization's well-known fax number on the top right-hand corner. France, when called upon by Japan to provide the original document in French, was unable to produce it (interview with Peter Bridgewater). What is interesting in this episode is that the way in which France conceived of buying back its green credentials was to perform the state's ability to "be democratic," to stand for "what the people of the world want for whales." In this move constructing the state's greenness and its democraticness thus appear as two sides of the same anti-whaling coin. It is also noteworthy that the lack of democratic transparency of some of these NGOs, whose IWC representatives are often self-appointed, appears to be of little consideration in adding this democratic strand to the construction of the ethical green state (for a broader critique of the democratic deficit of environmental NGOs, see Jameson and Eyerman 1989, Schwarz 1991, Spencer et al. 1991, Jordan and Maloney 1998).

The IWC annual Plenary has become a vast stage for the ritualistic conjuring of this anti-whaling "world opinion." The IWC meetings have historically been a site of increasing importance for anti-whaling NGOs: in chapter 7 we already saw the constant rise in NGO attendance, which reached forty in the run-up to the 1982 moratorium; in 1983 this number rose to fifty-one. Since the mid 1990s the number of NGOs has oscillated between ninety-one and one hundred one. Despite the formation of pro-whaling NGOs to counterbalance this influence (see chapter 11), these continue to remain largely outnumbered by anti-whaling NGOs. The presence of Japanese anti-whaling NGOs (Greenpeace Japan, the Iruka and Kujira (Dolphin and Whale) Action Network) is of great significance in being able to construct a multinational, multiethnic people of sorts. They represent the people of these whaling countries whose ethical concerns are ignored by their own states. During the voting sessions themselves, there is considerable to-and-fro between the tables of NGOs and states as commissioners often come to consult NGOs before casting their vote. This popular anti-whaling opinion performed at the IWC is then relayed back to the world via a global media, thereby closing the circuit generating this "world opinion." This constructed world public opinion, in turn, substantiates the legal case for proclaiming whales' universal right to life. NGOs have come to constitute the frontline force of a new, universalizing order which is above all a *moral* order, that of a global green civil society. Thus anti-whaling NGOs exemplify a form of new "moral interventionism" exerting all manner of disciplinary effects for which NGOs at large have come under increasing criticism (Hardt and Negri 2001, Amoore and Langley 2004, Jaeger 2007).

Importantly, the credentials required to be able to thus speak for the world is simply to attend as an NGO—being able to prove one's legal status as an NGO and paying one's NGO attendance fee. It is not justified by reference to any surveys of what their membership want, nor even any country survey. In fact, the surveys that have been carried out have tended to show that, when consulted, people's attitudes toward whaling are actually more ambiguous than is often conveyed by the anti-whaling story-line upheld at the IWC. In 1992, the Gallup Organization undertook a six-country survey on "Public Attitudes to Whales and Whaling" that cut across both whaling (Japan, Norway) and anti-whaling states (Australia, England, Germany, and the United States). The questionnaire was designed to gauge how much people knew about whales and whaling and their corresponding reactions to specific policy options. The survey revealed the extent of the lack of popular knowledge about whales, which was strikingly at odds with the volume of popular discourse on the topic, and, perhaps less suprisingly, more pronounced in the anti-whaling countries. For example, the Australians, British, Germans, and Americans overwhelmingly believed (65–60 percent), incorrectly, that "all large whales species are currently in danger of extinction," compared with 50 percent in Japan and 41 percent in Norway (Freeman and Kellert 1992, 27). With regard to the minke whale, whose estimates of 761,000 had recently been made public, a majority of people in all countries except Norway gauged their numbers at under 10,000 (Freeman and Kellert 1992, 28). Even more remarkably, most people in *both* whaling and anti-whaling countries tended in fact to *agree* with the policy prescription of "only limited harvest from abundant non-endangered species," namely, that it was acceptable to harvest nonendangered species: agreement rates stood at 86.1 percent for Australia, 76 percent for the United Kingdom, 73.6 percent for Germany, 81.1 percent for Japan, and 90.1 percent for the United States. The lowest was in fact Norway, with a 65.9 percent agreement rate (Freeman and Kellert 1992, 25). The results of this public consultation add a particular twist to Keck and Sikkink's (1998) analysis of NGO influence, since, in the whaling case, it would seem to be the case that a key tool of such influence is *mis-information* rather than information politics, or maintaining people in ignorance of material facts.

Whither Other Green Issues?

In the absence of a whaling industry, taking a stance against whaling at the IWC is, for states, a relatively costless way of polishing one's green credentials, both domestically and within the society of states. The significations pinned onto the whale signifier evolved over time. The initial involvement stemmed from states' commitment to protecting endangered species. Over time, however, the issue became invested as the site for the construction of states' green identities. In critically unpacking a discourse, just as interesting as the articulations that have obtained are conceiving of some that have *not*, even if this remains by definition a speculative endeavor. For example—to dwell upon possible qualifiers of "greenness"—given that the anti-whaling discourse is articulated as a discourse on the global environment, it is interesting to note that the parallel between "being green" and "being advanced" or "civilized" has not carried over into, say, climate change discourses. In fact, as the anti-whaling states' positioning shifted to ethical grounds after 1992, the same year the UN Framework Convention on Climate Change was signed, there appeared to be almost an inverse relationship between states' loudness on the anti-whaling front and their degree of support for the cooperative efforts on climate change. Some of the most vocal anti-whaling states were also members of JUSCANZ (Japan, the United States, Canada, Australia, New Zealand), the hard-line, obstructionist grouping of developed countries that formed in the in the run-up to the 1997 Kyoto Protocol. That the correlation between "being anti-climate change negotiations"/"being barbarian" or even "immoral" simply did not obtain had more to do with the very different configuration of power relations in either discursive field than with which of the two positions (denying climate change or being anti-whaling) has more material consequences in terms of global environmental degradation.

Of course, the formation of JUSCANZ also shows the prevalence of economic interest in determining the ways in which states position themselves with regard to environmental issues, since it drew together some of the countries that were at loggerheads on whaling—Japan and Canada on the one hand, the United States, Australia, and New Zealand on the other.[4] The broader argument of this book, however, is that economic interests only partially explain state positionings on environmental issues. What the whaling issue reveals is that, in the absence of a clear-cut common economic interest, as that which was brought out in the climate negotiations (where the JUZCANZ group formed in reaction to the idea that developed countries would have to initially bear the brunt of emission reductions efforts), other *identity*-type interests may instead account for how states position themselves vis-à-vis specific issues. Conversely, it also illustrates how states can take on specific issues *in order to* define themselves in specific ways, and that, in constructing their identities, it matters little that some of the identities thus endorsed might be logically contradictory with other identities taken on elsewhere. What matters is which particular identity is performed in a given forum.

Constructing the Performant, Neoliberal State

Promoting Whale-Watching

Another strand in the anti-whaling state's positioning is the defense of the whale-watching industry, which flourishes in anti-whaling countries such as Australia, New Zealand, and the United States (see chapter 9). In other words, being anti-whaling is also a way of being pro-whale-watching; it serves to foster a new and booming market for the tourism industry. Of course, harnessing moral or cultural arguments to open up new markets is nothing new, and market discourses have always been civilizational discourses, the banner of civilization being often carried abroad along with the exported goods (see, for example, Gill 2003). A similar moral interventionism transpires in the anti-whaling states' dealings with their whaling counterparts. For example, the following exchange took place between Australia and Japan in 1989 on the floor of the IWC plenary:

> Australia referred to its own experience following the closure of its last whaling station and reported that some unemployment had resulted. Subsequently new "whale watching" activities have developed in some parts of the country. Australia also noted that interest in non-consumptive uses of cetaceans were beginning to develop in Japan as exemplified by "Whaleland" in Ayukawa.
>
> Japan stated that although it does not oppose non-consumptive use of whales under the Australian policy, it is felt that encouragement of such policy should be confined to the 200 nautical mile zone of Australia (IWC 1989, 41).

The polished language of the Chairman's report barely conceals what is playing out as a confrontation between a dominant, interventionist discourse, and a defensive, dominated discourse. What is noteworthy here is not only that Australia sees it as its role to enlighten Japan on where its whale-related interests lie and how it should use whales. It is the way in which the broader anti-whaling discourse establishes the two forms of use as mutually exclusive—or rather, that whale-watching should preclude whaling. Interestingly, Australia's intervention simply frames out the fact that Ayukawa is one of Japan's four remaining whaling communities, and that a significant attraction of its "Whaleland" for tourists is the chance to eat whale meat (Institute for Cetacean Research 1988b, 101).

In actuality, whaling and whale-watching coexist in many parts of the world, such as Norway, the Caribbean, Canada, and indeed Japan itself. Furthermore, a comparison with other examples of dual resource uses, such as forests, would suggest that the main concern in policy terms is not the types of use per se but rather whether they are carefully controlled. All forms of use can lead to a depletion of the resource, as, for example, when the volume of visits to fragile ecosystems are left unchecked. Equally with the whaling issue, whale-watching activities have been shown to negatively impact the whale stock's reproduction rates (Moyle and Evans 2001). In other words, the incompatibility between whaling and whale-watching is neither a practical reality nor a logical necessity. Yet it has very real practical effects, notably in terms of excluding the visit to local whaling communities as part of what constitutes "the tourist experience" in a country like Japan. This is especially noteworthy for an industry driven by the values of "authenticity" and the search for "local culture" (MacCannell 1999, Lanfant and Graeburn 1994). Setting whales apart from other natural resources is thus an effect of discursive domination, insofar as alternative practices in which they are envisaged *just as* any other natural resources were evacuated in constructing their special status in the above exchange on the floor of the IWC. Japan, in turn, adopts the only strategy afforded discourses of resistance, which is to reveal the arbitrariness of this "taken-for-granted" construct of the dominant discourse, to denounce them *as* effects of domination (Bourdieu 1983, 2001, 2002c).

Saving Whales because WE Don't Eat Them

The anti-whaling discourse does not merely create consumers for a new industry. Because of this constitutive binary (either whale-watching or whaling), it excludes an existing set of practices. And because the primary use of whales in whaling now is as a food, it is indissociable from a moral condemnation of specific food practices, that is to say, a culture (Douglas 2002). For Russell Barsh (2001) the whaling issue is the latest in a long line of imperialist practices of "food hegemony" that began with the Roman Empire's implanting its food production system throughout the lands that fell under its rule, extends via the European colonization and the supplanting of indigenous food systems with large-scale production of foods that either suited the European palate or nourished more effectively vast laboring populations in other colonies, and continues today with the dietary conversion of First Nations of the Arctic to a Western imported diet. Indeed, historically the ascendancy of one people over another has often involved the appropriation, not just of their territory, but of the power to redefine what they can and should eat. Reshaping food practices and redesigning taste has thus constituted a key dimension of the symbolic and cultural domination that operates at its keenest in the wake of political conquest. The claim here is not that anti-whaling states harbor some hidden agenda to take over whaling nations. Nonetheless it is interesting to note how some of the social and cultural processes that operate at their keenest in the wake of political conquest may also be at play—albeit more mutedly—even when territorial borders and interstate relations have long been settled. My concern here is to tease out the specific part played by states in the symbolic domination at work in processes that have caught the eye of anthropologists, sociologists, and historians, but more rarely those disciplines whose sights are set upon the state itself. It is, specifically, to capture how this role enters into their positioning on the whaling issue. That a state may have taken on the anti-whaling discourse to promote its tourism industry at home can be explained as part of a broader embrace of global market discourses. What I want to consider here specifically is how being anti-whaling is indissociable from meddling with food practices that take place not just at home, but in other states.

The Anti-Whaling Discourse as Governmental Discourse

One striking observation is that the more aggressive anti-whaling states are also the ones who have taken the tightest turn toward economic liberalization. In fact, the way states fall on either side of the anti-whaling/whaling divide in the IWC maps over almost exactly with the Organization of Economic Cooperation and Development's (OECD) categorization of the degree of liberalization of developed countries' economies. Taking, for example, the benchmarks on agriculture, Norway,

Japan, and Iceland are ranked among the least liberalized economies and the most interventionist states, with respectively 1.4, 1.4, and 1.6 percent of their gross domestic products (GDPs) earmarked as agricultural subsidies (OECD 2002a, 176). New Zealand, Australia, and the United States, on the other hand, are lauded for being "in line with the long term reform objective of reducing support to agriculture and eliminating market distortions," with support levels, respectively, at 0.3, 0.3, and 0.9 percent of their GDPs (OECD 2002a, 128).

From a discursive perspective, the issue is not assessing the degree to which these states have actually liberalized their economies but rather the way in which they deploy new forms of *noninterventionist* governmental practices to position themselves as successful players in a neoliberal free trade regime and in relation to its associated dominant discourse on globalization (see also Hay and Rosamond 2002, Diez 2001), which concerns the relational dimension of state identity construction. In chapter 4, I analyzed the ways in which states belong to particular social orders or fields when I examined the IWC as a society of whaling states. Here I envisage the global free trade regime as another social *field*, in the Bourdieusian sense of a structured space of differential positions ridden with competitive dynamics. The concern is to examine the ways in which these two fields of social interactions may overlap and how their overlapping contributes to changing the play of positioning with regard to whaling. The main insight drawn out by the Bourdieusian perspective was that the competitiveness between social actors works to uphold and reproduce the field, rather than tear it apart. This is thus a far cry from the take-as-much-as-you-can-and-run realist understanding of competitiveness that simply ignores that competition is a social phenomenon. Instead, being competitive is foremost about mastering the rules of the game, so as to better position oneself in relation to social others. "Winners" are those who have developed the most acute "feel for the game"; they have successfully internalized its discourses and norms (or *nomos*) and are best able to play it to their advantage.[5] In other words, "winning" is a function of successful socialization (see also Wheeler 2000). Hence it matters to states to be seen as competitive players. States who have chosen to cast themselves as "neoliberal" *care* about being ranked as "better performers" by the OECD, as standing on the right side of globalization, whether it is about attracting foreign direct investment (material benefits) or out of a broader concern about their reputation and credibility (symbolic benefits). The Bourdieusian perspective highlights the social, alongside the economic, benefits to being able to successfully cast oneself as a neoliberal state.

For these performant, neoliberal states, the game thus becomes how to uphold expanding, globalizing economies without intervening in them. Moral and cultural discourses are discourses about who people are and how they should do things, such as what they should visit as part of their tourist experience of a foreign country. Thus, for these states, speaking moral discourses can constitute indirect ways of reordaining entire fields of practice at no cost to their commitment to economic nonintervention *and* in a way that maximizes the nation's productive capacities. For example, Mark Laffey (1999) has analyzed the ways in which New Zealand—another state at the forefront of the anti-whaling campaign—developed a new identity discourse that "added an Asian strand" to a predominantly white identity as a way of cultivating an image of cutting-edge competitiveness and thereby integrated the country into the Asian boom. By investing in the terrains of culture and morality, neoliberal states have been able to substitute direct interventionism with new forms of *indirect* interventionism whose main modality is discursive. Supporting the uniformization of taste, by taking on moral discourses about what people should or should not eat, is consonant with the broader governmental strategies of these states, aimed at bringing about optimal productive conditions, that is, those that sustain the creation of new markets for national produce.[6] Thus, encouraging the convergence—or "burgerization" (Millstone and Lang 2003)—of all the diets across the globe onto the Western diet serves to sustain the producers of its staples, such as beef. Perhaps it is no accident that the major beef-exporting states (Australia, the United States, Brazil, and New Zealand) are also some of the most vocal anti-whaling countries. Because not eating a particular food can also be a way of encouraging the consumption of another—such as beef—insofar as it proscribes eating whale meat, the states' anti-whaling discourse appears in this light as one more governmental discourse in the arsenal of the neoliberal state.[7]

Conclusion: Reproducing the Anti-Whaling Discourse

This chapter unpacked the various story lines that constitute the anti-whaling discourse and how they changed over time. The initial narrative out of which the discourse spun was, whales should not be killed because they are endangered. From 1992 onwards, it became, whales should not be killed because they have a moral right to life. Furthermore, *we* like to watch them. Finally, whales should not be killed because we do not eat them. Together these different story-lines form the anti-whaling discourse, such that evoking one story-line implicitly conjures up the discourse as a whole (Hajer 1995, 62). Of course, not all the actors comprising the anti-whaling discourse coalition ascribe to all of these narratives. A key advantage of the story-line approach is that, in showing that these narratives are tied to the policy context rather than to the actors who take them up (Hajer 1995, 123), it shifts the focus to the discourses instead of the individual actors in the coalition. Recognizing the relative autonomy of discourses, it identifies their principle of coherence in the discourses themselves rather than in the speakers. In other words, what holds the disparate elements of the anti-whaling discourse together, and maintains it as a cohesive whole, are these story-lines, rather than the actors who take it on; such that the anti-whaling discourse persists as a maker of a single, albeit multifaceted identity, even while a particular story-line may fall to the wayside over time.

What the evolution of the anti-whaling discourse shows is that for states today being anti-whaling has little to do with the material reality of whale stocks. For these states, the IWC has become a stage for performing oneself as a "good" member of a society of states normatively ordained by the notion that killing whales is wrong (the anti-whaling *nomos*). Embracing the anti-whaling subject-position serves to cast oneself as an ethical, green, civilized, and democratic state, that listens to what The People want for whales. Moreover, it enables states to be "good" in yet another increasingly important sense in the society of states today, that is, as performant, neoliberal states. The anti-whaling discourse constitutes a governmental discourse that shapes whale-related productive practices around the world not by intervening directly but simply by dictating what are the "good" and "bad" ways of utilizing whales. The unwavering state interest in the IWC is thus symbolic rather than material, as are the benefits to be reaped from continuing to uphold the whaling moratorium and perpetuating the anti-whaling discourse. Their interest is in retaining what has become invested as a site for identity construction. It is also in retaining symbolic or discursive power—the power to ascribe meaning, and consequently to stake out practices, at home and abroad.

11

State Positionings (II): The Pro-Whaling Discourse

This chapter is the counterpart to the previous one. It unpacks and holds up the pro-whaling discourse as a foil to the anti-whaling discourse, in order to reveal what alternative significations continue to be framed out in the reproduction of that dominant discourse. The opposition between anti- and pro-whaling factions that is invariably reenacted every year at the IWC is a struggle over which meaning will be pinned onto the whale and, thus, how it can be utilised—it is a fight for discursive power. At stake is the power to draw categories, and decide what falls into them, and the power to define both "self" and "other." An important effect of the anti-whaling's hold over the IWC is that all the categories by which whaling is regulated—or proscribed—today, namely "aboriginal subsistence whaling" and "commercial whaling"—have been crafted by the anti-whaling discourse. This means that the whalers and their practices have been defined by those who not only do not whale but actively oppose it. What drives the pro-whaling reaction is reclaiming the power to define oneself. Consequently, the discourse it has yielded is no longer the same whaling discourse, carried over from the pre-1964 whaling order (see chapters 2–4). The anti-whaling discourse's decisive shift from a radical to a hegemonic discourse has inverted the power dynamics between the two discourses, such that the old dominant *whaling* discourse has now given way to a new, *pro-whaling* discourse of resistance articulated *against* the anti-whaling discourse. Moreover, that the symbolic power of self-definition is at stake explains why the pro-whaling coalition extends beyond whalers, to nonwhaling peoples and states. As a dominated discourse, the pro-whaling discourse, however, is neither as readily available nor as full-fledged as the anti-whaling discourse.

Thus instead of taking story-lines as my starting point, as these are less delineated in this discourse, I focus on key signifiers around which the pro-whaling discourse has crystallized at three different levels, namely, at the state level, the suprastate level, and the substate level. I begin at the state level with the national discourses of three whaling countries, Iceland, Norway, and Japan, and consider how whaling discourses enter into the construction of national identities. The second site of formation of the pro-whaling discourse is interstate discourses at the IWC. There it brings into play two, almost opposite, articulations of sovereignty around two nodal points, "food security" and "sustainable use." Lastly, "cultural identity" has provided the catalyst for a new pro-whaling discourse taking hold at the substate level. Its associated concept of "cultural diversity" serves, in turn, to link together these different levels, yielding a vast coalition of substate and state actors around the pro-whaling discourse.

Whaling and National Identities: Resistance at the State Level

This section examines the positioning of countries that define themselves as "whaling countries." These are not simply countries where whaling takes place, such as Canada, but rather countries where whaling is held up at the international level as a marker of national identity. In other words, a country that whales is not necessarily a "whaling country": the United States is especially useful for drawing the distinction between the fact of whaling and of its discursive identification. This chapter thus considers only those countries that have persistently cast themselves *as* whaling countries within a society of anti-whaling states. The whaling traditions of Iceland, Norway, and Japan are considered individually in order to analyze the specific ways in which practices occurring on the ground (or not) are mobilized to position oneself as a "whaling nation."

Iceland

Iceland presents a discrepancy between a loud whaling stance on the international scene and the whaling that actually takes place. Although Iceland initially lodged a reservation to the commercial whaling moratorium, it immediately relinquished it the following year and has since then, overall, toed the anti-whaling line. It suspended all commercial whaling in 1983, and by 1990 it had also abandoned scientific whaling, the only remaining legal form of whaling under the IWCR (Article VIII) in the absence of a reservation, and it awaited the results of the SC. In 1992 Iceland resigned and remained outside the IWC for a decade.

Yet over that period Iceland abstained from whaling. In 2001 Iceland rejoined the IWC, having this time lodged a moratorium reservation. At this point, Iceland experienced at its own expense the power of the anti-whaling discourse to prevail even over the normal application of international law—the provisions not only of the ICRW itself but those of the 1969 Vienna Convention on the Law of Treaties governing reservations. In an unprecedented move that took the IWC's own Secretariat by surprise, the anti-whaling countries mobilized its majority on the floor to strip Iceland of its voting powers that year, by calling at the last minute for a vote on Icelandic membership, an issue not normally subject to vote under IWC procedure (see chapter 4). Iceland's membership was officially restituted in 2002. Despite what was perceived as a slight to its sovereignty, Iceland continued to remain cautious in wielding its reservation, launching first a limited scientific whaling program in 2003 and resuming commercial whaling only in 2006, with a quota of thirty minke whales (out of a population of 43,000) and nine fins (out of 25,000). While highly symbolic as an act of resistance, in actual fact this hardly increased Icelandic whaling, for the Icelandic fleet only caught seven of each whale, and in 2007 the fisheries minister did not reissue a whaling quota on the grounds that whaling was unprofitable.

Yet for all its actual caution Iceland has never ceased to posture as a "whaling country." In fact, the world's hostility has tended to sharpen the pro-whaling attitude of the population. After Iceland suspended its commercial whaling, three out of four Icelanders remained in support of whaling (Brydon 1996, 26). During the 15 years when Iceland withheld from whaling, the shipping merchant Kristian Loftsson, nicknamed Iceland's Captain Ahab in the pro-whaling discourse, maintained his whaling fleet intact (four boats) in the hope that Icelandic whaling would one day resume. In Iceland the significance of whaling far outreaches its historic or economic importance: a single land-based enterprise was established just after independence from Denmark in 1944 to hunt the large fins and sei whales, and ten families scattered along the North Coast combined a seasonal hunting for minke whales with cod fishing as part of a local economy. Moreover, while Iceland derived 75 percent of its foreign export earnings from fish products, 1.3 percent only came from whale meat, sold mostly to Japan. For anthropologist Anne Brydon, the defense of whaling goes to the heart of Icelandic nationalism, envisaged as a "collection of discourses that takes as its object 'the nation'" (Brydon 1991, 61). Whaling stands at the confluence of three "spatial discourses" on property, territory, and nature. Its defense served to crystallize the resistance to external interference in a small nation that had only recently acceded to sovereignty and remained still largely exposed to the outside world. Whaling is thus not defended as a traditional activity, nor as a source of income, but rather as the sovereign right of the Icelandic people over their natural resources. It is noteworthy that the rejection of anti-whaling is couched as a discourse on nature: the anti-whaling stance is perceived as the sign of a dysfunctional relationship with nature bred in urbanized societies (Ris 1994). The deterritorialized, unbounded image of nature underlying "environmentalist" discourses is contrasted with a localized, firsthand experience of an "Icelandic" nature (Byrdon 1991, 360–362; Ásgrímsson 1997).[1]

Norway

Norway is the only IWC member country to fully whale today. It engages in "commercial whaling," in that whaling is an integral albeit minor part of Norwegian commerce, restricted to the domestic market as a result of the ban on the international trade of whale parts in place since 1983 under CITES. It also undertakes "scientific whaling"; Norway has been an important contributor to the work of the IWC's SC. Norway, however, *just whales*. It is the only country that has been able to outright reject the drawing of distinctions between types of whaling. Unlike Japan or Iceland, Norway has not sought to legitimize its whaling by using either the "aboriginal," "subsistence," or "scientific" labels. How does it maintain this stance?

Norway had initially lodged a reservation on the commercial whaling ban in 1982, thereby preserving its right to whale, and unlike Japan and Iceland, it never retracted it. However, that year it also suspended all whaling, redirecting its efforts instead to the scientific front. After the member states rejected the new management formula proposed by the IWC's own SC (the RMP), Norway announced that it was unilaterally resuming whaling, using the same RMP that the IWC itself proceeded to endorse the following year (after the SC Chairman had resigned). The hunt was restricted to one species only, the minke whale, whose abundance estimates had been successfully determined. Minke whale quotas are established each year using the CLA, the precautionary formula at the heart of the RMP, combined with large-scale cetacean sighting surveys (carried out in 1987, 1989, 1994, and 1995). For example, the quotas were set at 671 minkes for 1998, 753 for 1999, 655 for

2000, and 711 for 2003. In addition, Norway developed a DNA register system to ensure the traceability of all catches since 1997 (Raysmaker 2001). In January 2001, the government lifted its ban on the exports of minke whale products; the first exports of meat and blubber to Iceland and the Faeroe Islands took place the following year.[2]

In line with a long history of whaling regulations (see chapter 4), Norwegian whaling is tightly controlled. It is restricted to territorial waters and thus strictly kept under Norwegian law, and it is limited to short seasons opened and closed by the Fisheries Directorate. The quotas are shared among a small and tightly controlled number of whaling boats, thirty-six in 1999, thirty-two in 2001, and thirty-six in 2003. The vessels are relatively small (70 feet on average, 37 tonnes), thus effectively precluding the catching of any species larger than the minke, which is one of the smaller great whales. Licenses must be renewed every year. Every vessel is inspected before the start of the hunting season. The whaling gear consists of a harpoon with an explosive grenade. In addition, the whaleboat is required to host a government inspector for each expedition. The inspector (generally a veterinarian) records all the information related to the circumstance of the catches, in accordance with the guidelines of the IWC management schedule. In addition she or he collects the data needed for the establishment of the whale's DNA fingerprint.

Beyond whaling, Norway's attitude of prudent harvest and its stringent regulatory framework form part of a broader commitment to a rational, science-based utilization of its resources. Thus the whaling issue is not considered in isolation but rather it is framed within a broader discourse on the state's responsible attitude toward natural resources at large. In the words of former secretary of the Norwegian pro-whaling High North Alliance (HNA): "The issue is bigger than whales. It is a question of the principles of international environmental and resource policy, of respect for international agreements and of the terms for maintaining cultural diversity" (Blichfeldt 1994, 2).

Whales (and seals) are considered just one among many maritime food resources in a country centrally dependent on fish products (the world's second largest exporter and tenth producer of fish; OECD 2001), and whaling (and sealing) are regulated as subsectors of the fisheries. Moreover, whale meat, "the meat of the people from the North" who live in harsh climatic and geologic conditions and where agriculture has traditionally received high levels of support, does not benefit from the subsidies accorded to land-based meat production (chicken, pork, and lamb).

In the seafaring, barren regions of the High North, minke whaling is a key part of the life of local communities. It is a tradition perpetuated from father to son.[3] In the Lofoten Islands, minke whaling is an integral part of cod-fishing, the islands' main activity. There whaleboats are also cod-fishing boats, and the whalers–fishermen harvest cod for three-quarters of the year and whales for the remaining quarter, during the early summer weeks when the cods spawn in the lush waters of the Norwegian sea, thereby drawing in the minke whales for their feed. Both the whales and the whalers compete for cod: a feeding minke weighing on average 5 to 8 tonnes consumes roughly 5 percent of its body mass daily—an average estimate of 250 kilograms per animal per day. Thus harvesting minkes is seen as integral to the broader ecological balance of the area. During the whaling season, which lasts around four to six weeks, whalers typically set off on two whaling expeditions to catch their allocated quota of fifteen to twenty minkes. The hunt itself is highly challenging, requiring both skill and an optimal and rare set of conditions that obtain only for few days in May–June, namely, twenty-four hours of daylight and a perfectly flat sea—or else the fin of the whale is impossible to spot (interview with Jan-Odin Olavsen). Once the adequate prey has been identified (a minke that is neither too juvenile nor with calf), under the watchful eye of the government inspector, the whale must be carefully handled, for, swifter than the vessel, it escapes if it suspects that it is hunted (interview with Bjorn-Hugo Bendiksen). The grenade explodes upon contact to maximize the chance of instantaneous death. Moreover, while the harpooner shoots, another crew member aims at the animal with a rifle, again under the gaze of the government inspector, who records the animal's time to death. Each animal yields around half its size in whale meat, between 1.5 and 4 tonnes. The blubber, not consumed by Norwegians, is thrown back to the sea; some of it frozen, in anticipation of the resumption of the whale meat trade.[4]

Although essential in the North, whaling is not central to Norway's economy as a whole. Average consumption of whale meat in Norway amounts to 1 percent of annual consumption of seafood (Raysmaker 2001). Hence, in terms of state positioning, the material interests of a few have been moved to the heart of the nation as a whole. That is to say, the significance of whaling is symbolic. Norway's whaling stance conjoins particular articulations of sovereignty and cultural identity, respectively, centered around the use of natural resources. First, for a country with an also relatively recently acquired independence (see chapter 4) the right to whale is an expression of its sovereignty. Moreover, notwithstanding (or rather because of) the recent mineral-resources-powered prosperity, Norway has been careful to cultivate its image as a "green country" with stringent environmental policies. Indeed, insisting on its environment credentials forms part of its discursive strategy at the IWC.[5] For Norway, the responsible management of whaling balances out the country's sovereign rights over its natural resources with its environmental responsibilities, in accordance with Norway's obligations under international law. It is noteworthy that the Prime Minister who oversaw the resumption of whaling was none other than Gro Harlem Bruntland, who had also presided over the UN Commission attributed with having coined the concept of "sustainable development" (the 1987 *Bruntland Report*). In Bruntland's own words upon Norway's resumption of whaling:

> A great deal of misinformation is currently being spread about Norway and whales because we intend to resume minke whaling on a modest scale, catching a few hundred this year. Some participants in the campaign against us believe that whales should not be hunted regardless of scientific or ecological justification for doing so, or they choose to ignore the best science and argue that Norway is threatening the whales and breaking international law. Neither is true. . . .
>
> The scientific committee of the IWC has now unanimously concluded that the best estimate of the North East Atlantic stock of minke whales is 87,600 animals. This stock can easily sustain a modest harvest of a few hundred whales and still continue to increase. But the IWC plenary, a highly politicised body strongly influenced by animal welfare groups, failed to act upon the conclusions of the scientific committee (Bruntland 1993).

Second, whaling is defended as an expression of Norwegian particularism. Indeed, we saw the historical role of whaling in constitution of an independent Norway in chapter 4. Thus the defense of whaling is a defense of Norwegian identity. The year after resuming whaling, Norway, whose accession to the European Union had been on the table since 1972, voted *against* joining by popular referendum. My point here is not to reduce the rejection of EU membership to a single issue but rather to highlight a broader identity-defining moment where Norwegians were intent on asserting their *difference* rather than similarities with neighboring states, and whaling incarnated this difference. Also fueling the support of whaling is the sense of a broader misunderstanding of the country's relationship to "its" nature, or rather its sea (named indeed the "Norwegian" sea), and the way in which it plays into its identity as a maritime nation. In this sense, as well, it invokes a particular discourse on nature (Ris 1994, Christoffersen 1994, Kalland 1994b, 1994c).

Hence, in Norway, the opposite articulation obtains than that which is proposed by the anti-whaling discourse. The discourse on whaling is an environmental discourse if by that we mean a particular way of envisaging the relationship between humans and their natural environment. What both Norwegian and Icelandic discourses draw out is that, in order to constitute itself as the hegemonic discourse on the environment, the anti-whaling discourse evacuated alternative discourses on nature. Hearing these stifled discourses requires seeking out the whalers themselves. Upon my visit to Lofoten one such instance of alternative conception of nature was brought home to me by my conversation with Halvard Bendiksen. When I asked him to describe his practice, the harpooner fondly evoked "the whaler spot." This is the little white spot close to the animal's brain that provides the target for the aim. It is, in the whalers' words, "god's gift to the whalers," for it remains visible even when the animal dives, and hitting it provokes instantaneous death (interview with Halvard Bendiksen). The idea of a divine hand that facilitates the whaler's task would, of course, be pure heresy to the anti-whaling discourse.

Japan

Japan has not officially resumed commercial whaling. Japan's position is constrained by its having relinquished in 1987 under U.S. pressure the reservation it initially placed on the moratorium. Unlike Canada, which has permanently withdrawn and quietly resumed whaling, and Iceland, which withdrew and rejoined amid loud clamor but without resuming whaling, Japan has constantly remained within the IWC. Given that Japan has never exerted its option to withdraw its membership, and given that it has forsaken the use of its ultimate sovereignty card (its reservation), the determining factor in Japan's positioning at the IWC has in fact consistently been the concern to be seen as a cooperative, law-abiding member of the international society. These efforts have remained moot, because, I argue, they were inherently defeated by the strength of the anti-whaling *nomos*, since Japan has systematically remained the primary target of the anti-whaling discourses since the early 1970s (see chapter 8), irrespective of whether it is actually the country that whales the most. Consequently, what had begun in the early postmoratorium years as genuine attempts to use the IWC as a platform for explaining their whaling to the rest of the world progressively evolved into a much more assertive, provocative positioning, encapsulated by Commissioner Masayuki Komatsu's quip, which he repeatedly delivered to the Western press at the IWC in 2001 and again in 2003: "When we see whales, we get hungry." Japan thus changed from a "whaling" to an "anti-anti-whaling" diplomacy (Ishii and Okubo 2007).[6]

With the chances of overturning the whaling ban becoming increasingly remote, Japan has sought instead to obtain recognition for the "distinctiveness" of its whaling. The strategy deployed has consisted in trying to break open the binary classification underpinning IWC regulations of all whaling as either "aboriginal subsistence" or "commercial," by offering a third category better suited to Japanese whaling practices, "Small Type Coastal Whaling." The category was first put forward for consideration before the Technical Sub-Committee on Aboriginal Subsistence Whaling in 1986, the final year of phasing out of whaling operations under IWC legislation, where it was rejected. Since then, together with scientific whaling, it has become the line of defense for Japanese whaling at the IWC, and every year Japan invariably tables it as an agenda item, to no avail. Resistance thus crystallized around this alternative category to the one imposed by the hegemonic discourse, one that told a different story about whaling.

Like Norway and Iceland, Japan has traditionally turned to the sea, and there too whales belong to a natural environment customarily exploited by humans. As in Norway, the national position is built on very localized practice. It is articulated as a story-line about the defense of traditions and cultural identity. When Japan first proposed the new discursive category in 1986, it rested its case in the following terms: "The Japanese have a long-standing history with whales. Traditions and culture on whales and whaling, fostered over the course of history, have been handed down from generation to generation" (ICR 1996a).

The Japanese whaling tradition is multilayered; according to whaling historian Junichi Takahashi (2003), it blends an ancient local coastal tradition with a more recent national culture. In the four coastal towns where whaling still occurs today, Taiji (in the Wakayama Prefecture), Abashiri (in Hokkaido), Ayukawa (in Miyagi), and Wadaura (in Chiba), whaling is an ancient tradition dating back to the late sixteenth century (ICR 1988a). These four towns alone form the basis for the Small Type

Coastal Whaling category. The whalers of these towns are no more "aboriginal" than the rest of the Japanese population, yet whaling is no less central to the "subsistence" of the traditional life of these villages than it is to that of the Alaskan Inuit. Whaling is integrated into a complex system of social exchanges and religious rituals. In a Shintoist cosmic system founded on the interdependence of animal and human realms, and where plants, animals, objects are endowed with a soul, humans become indebted to nature when using its resources, a debt which is then honored through religious ceremonies (Kalland 1995). In whaling villages, annual religious rites performed in the temple celebrate the "souls" of the taken whale, each consecrated with a "Buddhist name" and its own shrine. Memorial tablets are commissioned at considerable expense by whaling companies and sanctified by a priest. The hunt is dramatized in "whaling festivals" (ICR 1988c). Furthermore, while commercial transactions *are* involved in traditional whaling, it remains a far cry from the for-profit, expansionist whaling widespread in the West, which was originally the type of whaling struck down by the label "commercial whaling." Anthropologist Milton Freeman (2001) evokes the IWC's "money fetish," by which monetary exchanges are seen as "bad" and where whaling is tolerated (under the aboriginal subsistence category) as long as it is for "subsistence" only and not for money. He draws out the cultural relativity of this largely Western conception of money, which ill translates to societies such as Japan. There, money is the required gift to be offered to the gods (*osaisen*) and at solemn events such as funerals (Freeman 2001, 132). In Japanese coastal communities, a complicated set of exchanges and obligations, comprising gifts and offerings, both in cash and kind, revolves around each whaling expedition. Hence whaling weaves a web of social relations fostered around the "symbolic debt," creating "long term transactions" that serve to reproduce both the social and cosmic orders (Moeran 1992). The consumption of whale meat forms part of a complex culinary regime (ICR 1988g). Each part of the whale is cooked according to specific rules and has its particular place in the composition of the traditional meal (ICR 1988d, 1996b).

The postwar period saw the development of a national whaling culture. Each fall the departures of the Antarctic whaling expeditions were celebrated as national events, and the whalers themselves became the "new patriots" of a vanquished nation, not least because they brought back huge quantities of meat in times of food shortages (Takahashi 2003, 63). The expeditions themselves were followed "like naval campaigns during the war"; whaling constituted a mediated experience shared by the nation as a whole, and one central to its (re)constitution as an imagined community (Anderson 1983). This media coverage brewed a nationwide culture, sustained by the widespread presence of cheap whale meat in schools, for example. However, according to Takahashi (2003, 69), a decisive factor in shaping these operations into a national culture was the pressure brought to bear on Japanese whaling by the world: "when Japan faced a real crisis for its existence ... Japanese people began to see whaling as part of their "cultural" traditions that needed to be saved and revitalized by conscious efforts." Thus national pride—the very motive identified by the CIA to explain the United Kingdom's postwar involvement in the Antarctic (see chapter 4)—is what is at stake in the expensive efforts deployed by the government to maintain the Japanese whaling tradition in the only category allowed under its IWC obligations (ICRW article VIII), namely, "scientific whaling."

Where does Japanese whale meat come from today? The largest portion of whale produce available comes from its JARPA scientific whaling programs (approximately 2,000 tonnes of whale meat from around 400 whales; Isihara and Yoshii 2000, 2), which are recognized by IWC scientists as having made some of the most significant contributions to the work of the SC (interview with Ray Gambell). In accordance with the provisions of ICRW Article VIII.2, which stipulates that the proceeds of the whales thus caught should not be wasted, the meat and blubber are sold on the Japanese market. The produce from the scientific whaling programs is distributed throughout the various markets through a centralized system, based on the statistics evaluating the rates of consumption during the five years preceding the implementation of the moratorium, and the sales, in turn, provide subsidies for the programs (Misaki 1996). Small Type Coastal Whaling represents a much smaller supply (around 300 tonnes). Drive fisheries (mainly in Wakayama) and hand harpoon fisheries (mainly Iwate Prefecture) together supply around 700–1,000 tonnes of product mainly from dolphins and some small whales (pilot whales, for example), which do not fall under IWC jurisdiction (Isihara and Yoshii 2000, 3). There are also some remaining supplies of frozen whale meat dating back to 1991, the year Iceland withdrew (around 1,100 tonnes left in 1997; Isihara and Yoshii 2000, 4).

After World War II, whale meat comprised 45 percent of meat consumed, declining to around 30 percent as consumption of other meats rose. Today, however, whale meat has become an expensive delicacy in

Japan, and it is no longer widely consumed. The Japanese consume around 60 to 70 kilograms of fish products per person each year. By comparison, in 1985, the last year whale meat figured in official statistics, per capita consumption was at 0.3 kilograms, down from 0.9 kilograms in 1975. The figure was estimated at 0.1 at the turn of the century (Whale Conservation Network 2001). What, then, of "popular opinion" regarding whaling? Over the last two decades, the construction of Japanese opinion has been a vast battlefield, pitting foreign anti-whaling elements and domestic pro-whaling factions, composed in either case of both states *and* NGO elements. For example, the anti-whaling Commissioner from Monaco, Frédéric Briand, explained to me in an interview that the "play of Japanese opinion" was one of the many strategic options considered by the anti-whaling state–NGO alliance to discredit Japan (another was to threaten Japanese scientists with international isolation). He recounted an episode where, in 1998, he introduced the question of toxins in whale meat (PCBs) to the agenda of the IWC, simultaneously alerting the World Health Organization, as a deliberate attempt to create an anti-whaling backlash in Japan "from the ground up." This led to the mobilization by consumers' unions and created a rift in the Japanese camp, where the health minister was pitted against the minister for fisheries (interview with Frédéric Briand). Japanese opinion has been cast in all possible directions. One poll conducted for Prime Minister Junchiro Koizumi's Cabinet prior to the 2002 IWC annual meeting in Shimonseki established that 75.4 percent of the Japanese supported whaling (Greimel 2002). The poll is not read here in terms of the veracity of the results but as the government's attempt to gauge where its "public opinion" lies.[7] It was also in response to the fact that the previous poll of Japanese people had been conducted by a foreign NGO, the U.S.-based IFAW, according to which one in ten Japanese supported whale hunting. Unlike the Norwegians or Icelanders, the Japanese appear significantly divided at IWC meetings. The NGO section is packed with both pro-whaling associations and anti-whaling Japanese NGOs, such as Japan Whale Conservation Network, Iruka and Kujira Action Network, or Greenpeace Japan. As whaling represents some of the more traditionalist elements of Japanese culture, it is not unanimously accepted, in particular among the younger generations. For them, whaling is arguably less a priority than belonging to the "global village," where being anti-whaling is the norm (see chapter 9). "Just like everyone else," these Japanese anti-whaling activists see whales and dolphins as *kuwai* or "cute" (interviews with Mikiko Hagiwara, Tomoko Kajiki, Mitsuru Naito, Ayako Hasegawa, and Nanami Kurasawa).

Articulations of Sovereignty: Resistance at the Interstate Level

On the one hand, the anti-whaling discourse is taking shape as a governmental discourse that helps states better position themselves in the neoliberal game (see chapter 10). How, then, have states who support whaling responded to this particular articulation, a key strength of the anti-whaling discourse? Far from drawing walls around whaling so as to maintain it as a separate issue altogether, these states have instead embraced the anti-whaling states' own strategy of inscribing their discourses within broader metanarratives and drawing on the play of positionings that occurs in other discursive fields, using the concepts of "food autonomy" and "sustainable use." First mobilized by what had begun as a reactive strategy, these nodal points have become key to catalyzing the development of full-fledged story-lines that have served, in turn, to draw an increasing number of nonwhaling states into the pro whaling discourse coalition.

Food Security

Coined within the development discourse and associated with institutions such as the FAO (Maxwell 1996), "food security" provides an interesting example of how concepts travel from one institutional context to another, thereby connecting different discursive fields (Hajer 1995). It was ushered into the neoliberal trade debate at the 1999 round of World Trade Organization negotiations in Seattle, not by developing but by developed "net importing" states who were also whaling states, namely, Norway and Japan (in addition to Switzerland). They defended their continued involvement in their national production and markets in the name of a tradition of "self-sufficiency" and the necessity to guard against "world market volatility." They sought to obtain "food security" considerations included among the GATT Article XX exceptions to free trade (FAO 2001, 54–57). While this was rejected in the collective negotiations, "food security" was nonetheless maintained as their individual line of defense. Norway was especially vocal in voicing "the need to protect local agricultural practises and biodiversity" (FAO 2001, 57). Japan for its part had long since articulated its resistance to the demands to open up its markets under the previous round of trade negotiations around the notion of "food security," pinned onto the highly symbolic rice crop (Francks, Boestel, and Kim 1999). It had notably hosted the FAO conference on Food Security in 1995 just as the Uruguay round was being brought to a close.

The "food security" story-line that these "low performers," by the OECD's benchmarks on agriculture, mobilize to justify their deviations from the prescriptions of the hegemonic discourse brings into play a conception of the state as guardian of the nation's self-sufficiency, which is being starkly drawn out by being attached to the symbolically and culturally charged realm of food production. Thus it invokes a particular articulation of sovereignty, grounded in notions of state autonomy and territorial integrity. The resistance to the neoliberal prescriptions of economic laissez-faire is pinned onto sovereignty, upheld to defend a different type of relationship between the state and its economy. In other words, sovereignty is held up as a shield to neoliberal governmentality. What these states are implicitly tapping into in evoking the "food security" narrative is the governmentalization story, according to which sovereignty is becoming increasingly undermined by the ongoing expansion of governmental processes to the extent that, at one point along the trajectory, as Foucault pointed out (Foucault 2003, 37–40; see also Epstein 2007), sovereignty itself begins to function as a pole of resistance to governmentality.

It is noteworthy that on the floor of the IWC, Japan regularly refers to the 1995 Kyoto FAO Declaration on Food Security in defending its whaling. For example, Japan evoked it in the debate about its scientific whaling permits at the IWC meeting in 2000. Both China and St. Lucia rallied to Japan's support, explicitly in the name of food security (IWC52 2000). More broadly, "food security" was named as one of the key rallying themes at a 2007 conference of pro-whaling states (IWC/59/7 2007). Yet the 2,000 tonnes of whale meat yielded annually by these scientific whaling programs are unlikely to alleviate the nutritional needs of the world's largest fish-consuming nation of close to 130 million. The reference to food security is not intended as a practical measure but rather as a discursive strategy to hook the defense of whaling to the broader and highly sensitive debate on food production in Western countries. It is an attempt to invoke the same articulation of sovereignty as that which is at work in the free-trade regime, where it serves to justify the resistance against the "push" of market liberalization. This is all the more interesting as Japan, in relinquishing its reservation, has effectively disavailed itself of its sovereignty card at the IWC and thus of the possibility of drawing on it to justify its whaling. The "food security" is thus an attempt to mobilize sovereignty indirectly by conjuring up the work it is accomplishing in another discursive field.

The "food security" story-line has orchestrated the convergence of developing countries toward the pro-whaling line. It has served to articulate the repositioning of the same developing countries that had been initially roped in by the anti-whaling NGOs in the early 1980s to secure the whaling moratorium (see chapter 7). Indeed, one of the most spectacular reshufflings of votes in the IWC has been the rallying of *all* the Caribbean states to the pro-whaling side: Antigua and Barbuda, St. Vincent and the Grenadines, Dominica (who withdrew in 1983 and rejoined in 1992), St. Kitts and Nevis, and Grenada. Panama, for its part, the first Caribbean state whose delegation was used to table an anti-whaling NGO proposal (see chapter 7), had resigned from the IWC in 1980 in reaction to the instrumentalization of its sovereign vote by Western NGOs. In its reaccession speech in 2001 it explicitly tagged its membership to the principle of food security and proceeded to vote with the whaling countries (IWC 2001). In this context "food security" takes on additional connotations as resistance to foreign interference. It taps into another, postcolonial articulation of sovereignty as both sovereignty over one's natural resources and the right of developing countries to choose their own path toward development. In their final statement, the 2007 conference of pro-whaling states for their part declared. "developed countries are oppressing every aspect of resource use and economic activity of small island coastal states including conch, sharks, tuna, small cetaceans and turtles. This is discriminatory" (IWC/59/7 2007).

The rallying of these Caribbean states, often prized tourist destinations, to the pro-whaling stance generated the "Japanese vote-buying" strand in the anti-whaling story-line (*Eco* 2001a, 2001b). The concern in a discourse perspective is not to verify the story-line's truth content but rather to observe how it is constructed and mobilized in the play of positionings at the IWC.[8] Not only does the story-line effectuate an exact inversion of a scheme that was devised by the anti-whaling NGOs themselves (see chapter 7) but it serves to cast doubt on the ability of developing countries to make sovereign, autonomous decisions, on par with the anti-whaling states. The implication is that these countries are incapable of coming to their own position on the issue, let alone one in favor of whaling, without being bought. That sovereignty was indeed what was perceived to be at stake was brought home by an episode of the 2003 meeting deliberations. After yet another incident where, prior to a significant vote, a Commissioner for a Caribbean state had been approached by an NGO activist who threatened him with a tourist boycott if he voted on the pro-whaling side, the IWC chairman took the unusual step of condemning these NGOs for their allegations, stating that they constituted an "insult to the sovereignty of these countries." Interestingly, this anti-whaling vote-buying narrative, far from turning small developing countries away from the whaling countries, has in fact reinforced their support, by adding the sovereignty string to their pro-whaling bow. Thus to Australia's criticism of its 2005 vote the Pacific island state of Nauru reacted by exposing it as an undermining of its sovereignty (Dorney 2005).

Sustainable Use

The "sustainable use" nodal point effects exactly the opposite strategy of inscription than "food security." It takes up the anti-whaling discourse on its own grounds, by drawing on *another* governmental discourse, that of sustainable development (Luke 1995, 1999; Kuehls 1998).[9] Hence it seeks to legitimize whaling by inscribing it *within* neoliberal processes rather than shielding it by recourse to sovereignty. Or rather, sovereignty is still invoked, but a completely different articulation thereof, and one that concurs with the neoliberal discourse on growth. At work here is the Principle of Permanent Sovereignty Over Natural Resources, ushered into international environmental law by Principle 21 of the 1972 Stockholm declaration and entrenched in Rio by the 1992 UNCED (Perrez 1996). This is the very same discourse that yielded the concept of "sustainable development" a decade latter. In other words, sovereignty is germane to sustainable development, and the sustainable development discourse can function both as a governmental *and* as a sovereignty discourse. These almost opposite mobilizations of the very same concept, while they may be dismissed as mere inconsistency from a logical perspective, are interesting from a discursive perspective. They reveal "sovereignty" as being neither a logical nor a real "thing," endowed with essential qualities and governed by the laws of noncontradiction, but rather a time-bound social construct (Bartelson 1995). Sovereignty is also, as Alexander Wendt (1992) famously put it, what states make of it.

If some whales are no longer endangered, this state pro-whaling storyline goes, then a carefully regulated harvest of these species constitutes a sound example of sustainable use of its resources by a state. We have already encountered this story-line at work in the Norwegian context. It also served to crystallize the growing support of developing countries with no prior involvement in whaling. Apart from the Caribbean states, the other major development in IWC voting patterns is the rallying of nonwhaling developing countries to the pro-whaling side. A first wave of such countries (Seychelles, Kenya, Egypt, Jamaica, Mauritius, Costa Rica) had joined the IWC in the early 1980s to support the anti-whaling moratorium. However, after 1986 they rapidly lost interest, many of them withdrawing from the IWC (Jamaica in 1984, Mauritius in 1988, Egypt in 1989, Seychelles in 1995) or simply omitting to renew their membership fees (Kenya, Costa Rica). The early 2000s saw a second wave of accession, this time on the pro-whaling side, including countries who were rejoining, after a period of withdrawal, to shift sides (Belize, the Solomon Islands, Senegal). Thus, at the 2003 IWC meeting, the following countries voted on the pro-whaling side (in addition to the above-mentioned Caribbean states): Belize (member since 2003), Nicaragua (since 2003), Mauritania (2003), Benin (2002), Gabon (2002), Mongolia (2002), Palau (2002), Republic of Guinea (2000), Côte d'Ivoire (expressed its voting intentions, then joined in 2004), and Morocco (2001). Furthermore, some of those who initially joined in the 1980s to vote for the moratorium have actually realigned with whaling countries (Senegal, Panama, Belize), sometimes after a period of withdrawal. These countries' pro-whaling positioning stems not from a material interest in whaling (since no whaling actually occurs in many of them) but rather from a principled stance for the right to use their own resources to develop sustainably. In its accession speech that year Morocco declared that it had decided to join "to contribute to sustainable development" as did Nicaragua (IWC 2003). At a separate press conference that year, the West African countries framed their involvement in terms of a broader interest in promoting the sustainable management of marine resources.

Cultural Identity: Resistance at the Substate Level

"Sustainable use" has also provided the nodal point connecting the pro whaling state discourse to alternative discourses on whales located *within* the state. In New Zealand Sir Tipene O'Regan opened a meeting of Maori tribes (*hui a iwi*) where the government's anti-whaling policies were tabled for discussion with the following words: "The

Maori conservation ethic is about sustainable utilisation and not ecological absolutism as promoted by Western colonialists" (Te Ohu Kai Moana 1997). His statement points to the two strands conjoined in the pro-whaling stance thus taking shape in this discourse on indigenous identity. Whaling is held up as an example of local, sustainable practice founded in traditional knowledge and as a bastion against the cultural imperialism of the postcolonial state. The resistance to New Zealand's anti-whaling positioning mounted by the Maori around the Waitangi Fisheries Commission in defense of their tribal rights is worth analyzing at length since it serves to unravel strands of the pro-whaling narrative spun from an identity that is defined *away* from the state rather than within it. It draws dynamics of resistance operating at the substate, as well as interstate, level.

New Zealand and the Maori

In this pro-whaling story-line the misrecognition of aboriginal rights and of the inherently sustainable nature of indigenous practices are tied together. The Treaty of Waitangi Fisheries Commission/Te Ohu Kai Moana was established in 1992 at the outcome of the final settlement that sealed the last of the fisheries claims put forward on behalf of the Maori people under the Waitangi treaty (see chapter 7). It is the guarantor of Maori fishing rights. For a culture where whales and fish are considered genealogically equivalent (*whakapapa*), the commission defending traditional fishing rights was deemed the adequate body for developing a Maori position regarding whaling policies. It was officially mandated to do so by a "meeting of the tribes" (*hui a iwi*) on June 11–12, 1997, in Kaikoura (Te Ohu Kai Moana 1997). In what was effectively the first act of its breakaway from the government, the previous year the Waitangi Commission had thrown its weight behind the Makah's request for an "aboriginal subsistence quota" at the IWC, thereby sharply demarcating itself from the position of the chairman of the working group on Aboriginal Subsistence Whaling, who was that year the New Zealand Commissioner himself.[10] The Waitangi Commission even proposed other nominees for the chair. In his letter to the Chairman of the Makah Tribal Council the Waitangi Commissioner articulated the Maori support in the following terms:

> We are advised that the nominated Chair for that position and his advisors have in the past reportedly vilified the customary practises of indigenous peoples of sustainably utilising renewable resources within their traditional territories.

Thereafter the Waitangi Commission became progressively more assertive: whereas it had officially supported the 1994 Southern Ocean Sanctuary defended by the New Zealand government at the IWC (Te Ohu Kai Moana 1996), in 2001 they rejected the South Pacific Sanctuary scheduled to be tabled by New Zealand and Australia at the IWC that year. In fact, the Commission:

> Request[ed] that New Zealand withdraw its proposal for a whale sanctuary in the South Pacific until such time as [a full consultation with the Maori] has occurred and the extent to which it impacts on Indigenous Peoples of the South Pacific has been fully explored (Te Ohu Kai Moana 2001, 6).

This jars with the joint statements by the governments of New Zealand and Australia in tabling their South Pacific Sanctuary proposal at the IWC meeting that they "have consulted extensively with the countries of the South Pacific"—unless that only applied to other countries rather than their own (IWC/53/18 2001, 4).

This stark internal divide was thus simply evacuated in the construction of New Zealand's position at the international level. In 2001 this was facilitated by the Maori ancestry of former Conservation Minister Sandra Lee, who opened her intervention at the IWC that year with a few words in Maori, true to the bicultural identity of the nation, but thereby also performing a national unity on the issue at the international level that did not exist. Moreover, New Zealand accompanied its South Pacific Sanctuary proposal in 2001 with a singing and dancing performance by indigenous people in their traditional attire to mark the support of Pacific indigenous populations for the sanctuary. Only the dancers and singers were not from New Zealand but from the sovereign Kingdom of Tonga, nor did the New Zealand delegation include any Maori chief (interview with Sean Kerins). The performance had simply substituted one indigenous people for another who belonged to a different country altogether, before an international audience vastly ignorant of these differences, nor were these specified. It was also, however, a diplomatic move vis-à-vis Tonga, which at the time was not a member of the IWC and was considering resuming whaling ('Akau'ola 1999).

As with the pro-whaling developing countries, the Maori opposition to New Zealand's anti-whaling positioning is a principled stance in favor of the right to self-determination and practical autonomy rather than grounded in any material interest in whaling. In fact, economically, the Maori are heavily involved in whale-watching. As Sir Tipene O'Regan himself underlined (1996), "Currently Maori have no intention to harvest whale species, what we are aware of, but rather have developed commercial enterprises based on the protection of whales." Moreover, the Maori do not kill whales; they have traditionally been "passive harvesters" of beached whales for bone and food (Te Ohu Kai Moana 2000). What they are defending is the right to envisage whales outside the categories imposed by the dominant discourse, within those conveyed by their own culture, and to use it accordingly. Hence even if their whale-related practices align with those prescribed by the dominant discourse, they are protecting the right to retain the decision as to how to use the resource. In other words, they are struggling for symbolic power, without which self-determination is but a dead letter. For them, full recognition of their cultural identity entails recognizing their ability to define themselves and their relationship to their environment. Self-determination and self-definition go hand in hand. At stake in the Maori's pro-whaling discourse of resistance is thus the reappropriation of the power to draw categories and to decree what falls into them.

The Maori's relationship with whales is ancient and mythological (see chapter 7). Whales are thought of as somewhat supernatural, a gift from *Tangaroa* (god of a sea) and the guardians of long ocean voyages (Te Ohu Kai Moana 2000). In some tales the first Maori arrived to *Aotearoa* on the back of a whale. In other versions a stranded sperm whale was the first treasure discovered on the new island and the first subject of dispute between the two original tribes (Te Ohu Kai Moana 1999). For the Maori, the whale thus ranks among the "treasured possessions" (*taonga*) for which they have been guaranteed "full and undisturbed possession" under Article II of the original 1840 Treaty of Waitangi (in return for the proclamation of British sovereignty over the Islands in Article I). In O'Regan's (1996) words, "we regard the interest in whales as a tribal property right." The opposition with the New Zealand government revolves around who has the power to classify whales. Denying the Maori recognition that whales fall into this category effectively maintains the resource under government rather than tribal jurisdiction. The 1978 MMPA outlawed the "take" of marine mammals in any form. The handling of whales, dead or alive, was made the preserve of the Department of Conservation. Each stranded whale is the object of expensive rescue operations involving Department of Conservation workers and volunteers, often anti-whaling activists, that meet with limited success, as the whales often return to where they were stranded, for reasons still little understood.

The Waitangi Fisheries Commission considers the Act in the following terms:

> While seeking to protect marine mammals, including stranded whales, the Act was another attempt to extinguish the Maori customary use of marine mammals, by making it illegal to use whales in all ways. Not only was access to meat, oil, bone and teeth lost, but the customary practises [which centered on the flensing process], or the primary vehicle for the transmission of traditional knowledge from generation to generation, were also severed (Te Ohu Kai Moana 1999, 4).

The stranded whale plays a distinct part in the Maori customary life. As Archie Taiaroa (1999), convener of the Maori Congress and member of the Te Ati a Paparangi tribe, explains:

> Each time a stranding occurred it provided the opportunity for traditional knowledge to be passed from generation to generation. Their place in the tribal *whakapapa* (genealogies), and details about their migrations, breeding habits and food sources were exchanged.

In response to Maori protestations, the Department of Conservation has accommodated for limited access to their "treasured possession" by setting up a permit system and a "bone bank" that metes out controlled amounts of whalebone for traditional carvings. This effectively maintains management of access under the control of the Department of Conservation, which retains discretionary power to withhold authorization over a permit. In practice, this has led to Maori being played off against each other, and those who oppose Department of Conservation policies have found it harder to obtain whalebone (interview with Sean Kerins). Furthermore, the bones are given out along with a good word of advice to switch to other types of bone (Te Ohu Kai Moana 1999, 4), illustrating once again the profound entwinement of discursive and moral categories. Once again, the bone of contention between the Maori and the government is the power of nomenclature (or naming) rather than the fact that the practice relates to a dead animal. Indeed, the Department of Conservation, also known as the "Department of Culling," readily deploys expensive culling programs to control the possums and stoats with killing methods (such as poison 1080, which causes failure of the internal organs; Paddock 2001) that, as pro-whalers point out, are far more inhumane, albeit less spectacular, than the explosive grenade (Sandoe 1994). The killing of animals—mammals even—is thus an integral part of resource management practice in New Zealand. The issue is rather who has the power to proclaim a mammal as "a pest" or as "cute."

Pro-Whaling Substate Networks

The Maori's newfound assertiveness on the whaling issue owed to global rather than domestic developments. In 2000 the Maori tribes hosted the third annual meeting of the World Council of Whalers (WCW), much to the dismay of Nelson's parliamentarian and former conservation minister Nick Smith, who declared it "as welcome as the Ku Klux Klan" (Christian 2000). Reaction to what the whalers saw as a misperception of their culture had begun in the early 1990s, and by the late 1990s it had developed into a full-fledged global network of whaling peoples. The HNA was the first proactive organization set up in 1991 explicitly to carry forth the voice of the whalers.[11] Funded by the Norwegian government and various fishing and whaling unions in Norway, Canada and Iceland, Greenland, and the Faeroe Islands, it is run by pro-whaling activists dedicated to breaking the anti-whaling NGOs monopoly over information and explaining "what whaling is really about," in the words of its former Secretary Rune Frovik (interview with Rune Frovik).

The WCW, for its part, was established in 1997 "to provide a collective informed voice for whaling peoples around the world" (WCW 2004). The WCW's annual General Assembly is a vast encounter of whaling traditions, both aboriginal and nonaboriginal. It offers a platform for whalers to be heard internationally, when domestic forums are barred. No doubt the support mustered by this nascent global pro-whaling network helped the Makah overcome virulent local reactions and finally set out to catch their one and only whale, three years after they had formally obtained the right to do so at the IWC.[12] The WCW is a forum for exchanging advice (notably on tribal rights and government relations) and formulating common policy positions through resolutions adopted at its General Assemblies, thereby performing, to paraphrase its logo, "the united voice of whaling people around the world." Some of its key positions include, first, the rejection of the IWC's distinction between "aboriginal subsistence" and "commercial whaling." In the WCW's (2004) own words:

> The World Council of Whalers respects the historic traditions of all societies that carry out whaling on a sustainable basis, and consequently the Council does not distinguish between commercial and non-commercial sustainable whaling (WCW 2004).

Second, the whalers themselves have pronounced unanimously against the resumption of large-scale industrial whaling (WCW 2000). Third, they have called for the recognition of the "sustainable use of non-endangered whales" (WCW 1998a).

In creating this transnational alliance of local cultures, the pro-whaling NGOs employ, to varying degrees of success, the same tools that were used to spread the anti-whaling discourse, namely, media visibility and long-distance communications that connect whalers from opposite sides of the globe. A crucial feature of the WCW is its Internet Web site, which offers itself as a window onto whaling around the world, both for the whalers to discover each other and for the rest of the world to discover whalers. Other developments include Internet "whaling libraries."[13]

"Localness" and "Community"

The recovery of such tools forms part of a broader strategy to reposition local whaling cultures in relation to the global anti-whaling discourse. As a result, the whaling identity itself has been transformed. In other words, it is not the taken-for-granted identity that obtained in the pre 1964 global whaling order (see chapters 2–4). It is a new identity, born out of resistance to the anti-whaling discourse, that is *also* grounded in sometimes ancient whaling practices. It is an activist identity, reclaimed against a dominant global discourse and very aware of itself as the identity of a dominated and resisting minority. In other words, the new pro-whaling identity is an anti-anti-whaling identity. Two signifiers are key to this repositioning, namely, "localness" and "community." Against the deterritorialized, individualized identity offered by the anti-whaling discourse (see chapters 7–9), the whalers constantly emphasize the *rootedness* and collective dimension of theirs. In this pro-whaling story-line, "community" is the place where a "real" culture sits, one deeply rooted in ancient traditions. The immediate interactions with, and intimate knowledge of, real whales is contrasted to the images of the virtual whale and to the mediated, far-removed, and largely ignorant relationship with nature conveyed by the global anti-whaling discourse. "Localness," in turn, is held up against the uniforming pull of globalization that threatens to eradicate cultural differences. "Local community" is reinvested as a site of resistance. Their strategy thus pertains to a new type of political practice that has taken shape in reaction to globalization that Arturo Escobar (2001) has identified as "strategies of localisation." Theirs is thus a "dialectic of localising and globalising" where, even while connecting remotely from different corners of the globe, they engage in "place-making practices" where the "local community" is upheld against the sense of dislocation and placelessness wrought by globalization (Escobar 2001, 146; see also Escobar 1996). The pro-whalers effectively position themselves as anti-globalization activists advocating cultural diversity.

Cultural Diversity: The Convergence of Substate and State Actors

The Right of Peoples to Choose Their Own Diets

The right to cultural self-determination, pinned onto the eating of whales, has provided a key nodal point for linking these substate storylines about localized identities back to interstate dynamics. Thus the "right of people to choose their own diets" is a key trope of the pro-whaling state discourse at the IWC (see, for example, IWC/59/7 2007). Just as the possibility of eating whale meat is upheld at the local level against the cultural domination vehicled by the state, it is upheld by states at the international level to resist the cultural homogenization wrought by economic globalization. For example, Tonga's considerations about resuming whaling were framed *both* in terms of "evidence of increasing abundance of whales in Tongan waters" ('Akau'ola 1999) and in reaction to New Zealand's using Tonga as a commercial dumping ground for its cheap mutton flaps, the rise in consumption of which had been linked to the accelerated spread of obesity, cardiovascular diseases, and type II diabetes among Tongans ('Akau'ola 1999, National Food and Nutrition Committee of Tonga 1999, Fiji South Pacific Consumer Protection Programme 1999). Proclaiming their right to be different, by eating differently, is thus what binds together these very different actors into a vast pro-whaling discourse coalition of state and substate actors located in other states.

Cultural Diversity

In turn, "cultural diversity" provides the umbrella concept connecting these pro-whaling identity discourses. The 2001 UNESCO Declaration on Cultural Diversity has become another recurrent reference in pro-whaling state discourses at the IWC (see, for example, IWC/59/7 2007). It also serves to hook this discourse to the highly symbolic 1992 UNCED process via the Convention on Biological Diversity, which recognizes the role of local communities and traditional knowledge in biodiversity conservation (Preamble and Article 10.c). It thus links into another global green metanarrative, as yet another way of positioning the pro-whaling discourse as a counterhegemonic environmental discourse.[14] Interestingly, the whaling issue thus comprises one of the few areas where states have become the vehicle for channeling back habitually marginalized local knowledges (Shiva 1998, Magnusson and Shaw 2003) into interstate discourse. Thus the North Atlantic Marine Mammal Commission (NAMMCO), a regional organization established in 1992 partly out of frustration with the policy gridlock at the IWC, seeks to integrate the methods of "traditional knowledge" to their resource management practices. In 2003 NAMMCO hosted its first Conference on User Knowledge and Scientific Knowledge in Management Decision-Making. It will be interesting to observe how the Declaration on the Rights of Indegenous People, adopted by the UN General Assembly in September 2007 despite the negative votes of Australia, New Zealand, Canada and the United States, plays into this cultural diversity story-line of the pro-whaling discourse.

Conclusion: Reclaiming the Power to Define Oneself

This chapter has analyzed the pro-whaling discourse of resistance focusing successively on the different levels at which these dynamics took hold: at the national level, at the interstate level, and at the substate level. At the national level, the pro-whaling discourse is a discourse about national identity. At the interstate level, it is a discourse about sovereignty and noninterference in the state's the use of its natural resources. At the substate level, it is a discourse about cultural identity and the right to self-determination. Central to the analysis were the key signifiers that function as nodal points linking up these various story-lines into a cohesive pro-whaling discourse and its associated coalition of state and nonstate actors.

Because it took shape against the dominant anti-whaling discourse—because it developed as an *anti-anti-whaling* discourse—and because it retains significations that were evacuated in the latter's rise to prominence, the pro-whaling discourse was central to drawing into relief the operation of discursive hegemony. It is, however, at the substate level that this operation is exposed at its clearest. The Maori, who do not kill whales, refuse their government's telling them they cannot do so. What unites the whalers' rejection of the distinction between "aboriginal subsistence" and "commercial" whaling, Japan's demand for the creation of a new whaling category (Small Type Coastal Whaling), and even Japan and Iceland's distortion of the category of "scientific whaling" is a fundamental rejection of discursive categories defined elsewhere and imposed from without. The Maori's pro-whaling positioning in particular forms part of a broader rejection of the notion of "subsistence," which is steeped in discourses about the intrinsic nature of indigenous population ("what they are really like") and from there, lays out neatly demarcated paths for their development, thereby effectively serving to reproduce these Orientalist stereotypes. Positing subsistence activities as "appropriate" for indigenous populations only serves to maintain them below levels of "real" (hear "commercial") development. It was on this basis that the 1892 Oysters Fisheries Act and the 1896 Sea Fisheries Amendment were overturned by the 1989 and 1992 Waitangi Fisheries settlement between the Maori and the New Zealand government, because they were designed to maintain Maori fishing "as subsistence only" and therefore denied "the right of the Iwi [tribes] to develop their economic base" (Te Ohu Kai Moana 2001). Similarly, Fabienne Bayet (1994) has shown how the assumption that ecotourist activities, such as whale-watching, are the forms of activity best suited for aboriginal people builds on stereotypes about their inherent connection to nature. It merely reproduces the nature–culture divide, a founding exclusionary binary within Western thought that also enabled the colonial subjugation of indigenous populations. The different claims contained in the various strands of the pro-whaling discourse are all, in one way or another, about reclaiming the power to define oneself. Hence it is not just about the right to use the resource according to their culture, nor just about sovereignty and noninterference, or about sustainability, or even just about the right to self-determination, even if it is also about all of these. For it is centrally about *self-definition*. What they are reclaiming is the power of nomenclature—the power to name oneself and the world around, and one's relationship to it.

12

Conclusion: The Study of Identity in International Relations

To conjure up a forgotten whaling world, this book began by mapping out the various forms of whaling around the world, past and present. Excavating this world from within our own past illustrated the extent to which whale parts were pervasive in the everyday life of the modern individual. Starting from the whaling practices on the ground served to demonstrate the unity of a social system where what is said about whales is tightly bound up with what is done to them and where individuals and states are connected within particular structures of normalization that, here, entrenched whaling.

Whales provided a key raw material, a fuel, and a food, and no doubt it was just as difficult to conceive of doing without whale products then as it is today to envisage doing without petroleum products. Chapter 2 then analyzed the structures of the modern whaling industry. Whaling evolved from a craft into a large-scale production system and became a typical heavy industry of the second industrial revolution. As such, it was essential to rapidly industrializing economies, and in some countries it came to incarnate the modernization process itself. This significance of whaling was acutely revealed by the wars of the twentieth century and by the rise of what I termed whaling nationalisms. It also explained how more and more countries became embroiled in an increasingly intense competition for dwindling whale stocks in a ruthless whaling Olympics that brought whaling to an end in the West. Yet the political economy of modern whaling revealed that the West's interest in whaling was not reducible to its material interests, since whaling lasted well beyond the point where it had become uneconomical and substitutes had been found for the main uses of whale parts. The West remained interested in whaling because its interest in whaling was shaped by an entrenched whaling discourse that lived on, increasingly disconnected from the materiality of whaling.

Chapter 3 continued to appraise the extent to which whaling was engrained in the structures underpinning this past whaling world by examining more closely the nexus of whaling and state practices. It shone the twin lenses of "sovereignty" and "governmentality" onto whaling practices to draw out the points at which whaling entered into the making of the modern state. Whaling, which was centrally involved in charting new waters and new territories, was key to drawing the contours of colonizing sovereignties and to marking the presence of maritime nations on the seas. For example, whaling was involved in organizing the disciplinary colonial penal system in the South Pacific, and it was implicated in trade wars that developed on both sides of the Atlantic with the rise of the state's governmentality functions. The importance of whaling framed the claims of sovereignty over Antarctica that were being put forward well into the twentieth century. Lastly, whaling's role in shaping the national interest surfaced in the particular structures of knowledge and power that yielded cetology, the science of whales.

The Social Construction of Agency

Proceeding further up the levels of analysis, chapter 4 analyzed the society of whaling states. It examined an international whaling discourse that crystallized around the development of whaling regulations, and it observed how these discursive interactions played into the making of states' whaling identities. Most centrally, what this sociological analysis drew out is that it is not simply that the world is socially constructed; *it is that the actors who construe it are socially constructed, too.*

Specifically, this chapter showed why analyzing the social processes through which actors' identities are construed mattered in the appraisal of the actors' interests. It began by opening up "interests" beyond the individualist, static, and essentializing understanding that has tended to prevail in the analysis of state interests by restoring its collective and dynamic aspects. States do not interact with one another from a set of pregiven and fixed interests and identities; rather, their interests and their identities are shaped by the discourses in which they are immersed and the fields of interactions in which they take part. Subsequently there is an interest, shared with other actors, in continuing to belong to a particular field of interactions, simply because it is where their interests take shape. Moreover, this field of interactions is more than the sum of its individual interactions; it is a *social field*, with its own normative order, or *nomos*, its own discursive regulatory mechanisms of recognition and shaming. The IWC was staked out as a social field, with differential positions of power, by specific dynamics of inclusion and exclusion. It was founded as a select, members-only club of whalers, not constitutionally but by the de facto exclusion of the second and third largest whaling nations (Japan and Germany), thereby effectively denying them the possibility of being recognized as whaling nations.

More than any material benefits, these dynamics of social belonging are key to explaining states' enduring interests in the IWC, both then and now. Then, the IWC provided a stage for members of the whalers' club to perform their interest in whaling and thereby express their identity as whaling nations, even after they stopped whaling. It thus enabled Western states to play a larger role in the international politics of whaling than their place in the trade actually warranted. Now, it offers states a stage for performing themselves as anti-whaling countries. While the nomos has shifted, the IWC has consistently remained the site for the construction of whale-related identities, which are, in turn, shaped by the way the nomos lies.

States' changing positions at the IWC are not reducible to a norm's being realigned along shifting material interests. This fails to explain why states continued to remain involved when there was no commercial interest in not whaling (prior to the economic takeoff of whale-watching). States could have simply withdrawn from the IWC altogether when they stopped whaling. Or alternatively they could have withdrawn when the new norm was at odds with their continued interest in whaling (some did). What explains social actors' investment in a norm is that it serves to define them, to position them in relation to the other social actors within the same field. In other words, understanding the dynamics of identity construction is key to understanding how norms actually work To appraise their functioning, this study has shifted the focus away from norms and onto the discourses that shape both the norms themselves and actors' identities. Its central implication is that the dominant discourses that regulate a particular issue–area of international relations are implicated in the constitution of state identities.

The Duality of Discourse Production

I then turned to considere the production of an anti-whaling order. In appraising the birth of a dominant anti-whaling discourse, chapter 5 explored the complex interplay between agency and discourse's productivity. Specifically, it showed how actors produced a discourse that, in turn, produced them as agents who occupy distinct and consequential subject-positions. While the anti-whaling discourse was enabled by the end of whaling in the West, these material conditions alone were insufficient to explain why a radically different discourse on whales actually took hold. The anti-whaling discourse was produced by a specific set of actors, environmental activists. This, in turn, was not sufficient to explain how it became a *hegemonic* discourse, such that it successfully eradicated not just the whaling discourse of the past whaling order but all other ways of seeing whales that obtained elsewhere. What produced it into dominance, as it were, was that it was coined at the juncture of, and thereby served to reinforce, two preexisting powerful discourses, an individualistic Cold War discourse on capitalism and democracy, and a nascent global environmental discourse. The dominant anti-whaling discourse was founded by a double synecdochic move, whereby increasing yet inchoate concerns about global ecological destruction were pinned onto the plight of the whales. Once dominant, however, the discourse, in turn, produced these environmental NGOs as powerful actors in the international system, to whom states listened. The new subject-position proposed by their discourse, the anti-whaling I/we, established these NGOs as full-fledged actors in the international system who wrote the policies of the IWC. This discourse thus brought nonstate actors a voice in a system dominated by states. This, in turn, helps explain the continued importance of this discourse to some international environmental NGOs today.

Knowledge or Beliefs?

Chapter 6 analyzed science's limited powers to influence policy making. On the one hand, science is discursively mobilized as the language of rationality and the authoritative discourse on truth. That is, we expect science to tell us how things really are out there. More importantly, we expect that the provision of perfect scientific knowledge will resolve all remaining disagreements, because it will unequivocally reveal the appropriate, rational course of action with regard to a natural resource and the environment at large—so much so that enduring policy differences are reduced to issues of imperfect knowledge and persisting problems of scientific uncertainty. Yet the story of science in the IWC is one of a science that is increasingly improving, yet decreasingly listened to. Instead, the trajectory of science on the whaling issue suggests that, were science able to eradicate uncertainty and provide us with perfect knowledge, such that it would be unambiguously clear what needs to be done, it is far from clear that we would be able to listen. Indeed, this capacity to hear, in turn, is determined by the subject-positions from which we listen. If science speaks to us in a discourse that is completely at odds with our own—with the one we have chosen to speak, because it marks us in certain ways—then it is likely that we will simply not listen. What science does not have is the power to alter a society's normative order. In fact, its own knowledge frames are informed by these underlying norms.

Chapter 7 examined the practices and strategies deployed by environmental NGOs to obtain the indefinite suspension of whaling at the IWC in 1982. The anti-whaling campaign unfolded on two successive levels. The subject-position carved out by the anti-whaling discourse was designed to interpellate individuals who, in stepping into it, marked themselves as caring for the whales and the environment at large. In other words, it crafted an identity at the individual level that was designed to transcend particularistic markers of identification, notably nationalistic attachments that could alternatively ground individuals' allegiance to their whaling states. At a time when whaling was petering out in the West, this proved extremely effective. In skillfully utilizing a nascent global media, anti-whaling NGOs successfully created a global constituency of anti-whalers, which then brought them the political clout to be taken seriously by states. At this point, the campaign turned to the international organization in charge of whaling. At the IWC they called upon states to step into the anti-whaling subject-position, but they did not stop at that strategy alone. There, in addition, *they themselves* stepped into the states' own institutionally determined subject-positions. They established themselves upon state delegations—both their own and those of other states—and in some cases, they literally appropriated states' voting powers by *passing for* states.

Theoretically, the chapter illustrated how this "stepping into," the process constitutive of actors' identities also comprises a precondition for action and policy making. In other words, social actors—whether individuals or states—engage in the realm of action by first stepping into subject-positions, that is, by taking on certain discourses. It is these discourses that make certain courses of action desirable and precludes others. Put differently, social actors see these courses of action as the right one to take from particular subject-positions carved out by particular discourses. Because it focuses on social actors as *speaking actors*, the discursive approach offers a way of bridging the different levels at which international policies are shaped. In other words, the analytical focus encompasses all the levels of analyses pertinent to the study of international relations, from NGOs to states.

The Role of "the Other" in the Construction of Identity

Having assessed how the anti-whaling discourse came about, in the chapters in part III, I examined the factors that have enabled it to persist. The effectiveness of the discourse itself proved a central factor. Hence having observed the political effects of this subject-position in chapter 7 and how it was successfully mobilized to effect change, chapter 8 examines how this subject-position itself was discursively construed through a detailed analysis of an anti-whaling text. This text offers a manner of a snapshot into a constantly circulating discourse where its salient features are temporarily arrested. Comparisons with an anti-whaling caricature currently in circulation on the Internet served to show the persisting resonance of the discourse today and the continued validity of its representations. The anti-whaling discourse owes its effectiveness to the particular representations of identity it conveys, that is, specific constructions of "us" and "them" that are implicitly reproduced every time the anti-whaling discourse is spoken. These were brought to light by the methods of discourse analysis.

The anti-whaling discourse posits the whales as passive, magical objects, victims of the cruelty of "them-whalers." This aggressive, bestial subject, in turn, conjures up a heroic subject to stand up to it, thereby carving out the subject-position for an "anti-whaling us." The anti-whaling discourse thus operates on the constitution of an "us" against a "them," a global civil society of anti-whalers united in opposition to the barbarian whalers, best represented by the Japanese. Interestingly, the discourse is not in keeping with the reality of who actually whales and, more significant still, with who has historically most depleted the whale stocks. The anti-whaling discourse thrives on its ability to evacuate the West's historical responsibility with regard to whaling and to the corresponding construction of a blameable "other." In other words, amnesia is a key mechanism of the anti-whaling discourse. What it does specifically is that it transforms an *antagonism* into an *opposition*. Because of the West's whaling past, the relationship between the Western anti-whalers and the whalers is inherently antagonistic. That is, it is the sort of relationship where the presence of the "other" threatens the neatly demarcated boundaries of the "self," because the "other" is too similar, not other enough, as it were. By setting up this clear-cut us/them binary, the anti-whaling discourse successfully "othered" the other. Obliterating any similarity with the self serves to leave the other standing separate and truly *other*.

This book foregrounded a particular modality of identity construction that has become increasingly salient in international politics, namely, the scheme "us versus them." The analysis of this particular us/them binary illuminated the processes underlying broader articulations of "them" against an "us." In particular, it reveals the central function of "the other" in the constitution of identity. The other serves to define the self by establishing *who it is not*. Exclusion and differentiation constitute key mechanisms of identity formation, and the other serves to shape the contours of the self. Once again, this draws into sharp relief the inherent unfixity underlying identities, which are always in the making and cannot be explained by reference to preexisting essences. Moreover, this study revealed the emphasis on difference as a mechanism for masking, and indeed performing away, latent similarities with "them," whoever "they" may be. It brought into perspective the profound similarity that drives the splitting of an "us" from a "them."

Why Identity Matters to the Study of Norms

Chapter 9 examined the consumptive practices that sustain the anti-whaling discourse. It showed how the anti whaling identity is construed not simply discursively but through concrete modalities of consumption. An overview of these immaterial forms of consumption, for which the whale must be kept alive and whole, mirrors the earlier survey of the very material consumption of sliced up whales that fueled whaling economies. The chapter then analyzed these forms of consumption against broader evolutions in our economies, highlighting, the shift to a service-based, increasingly "virtual" economy, the rising production of images, and the importance of tourism as a consumptive experience. For example, in donating to help save the whales, the individual is marking herself (or himself) as an anti-whaler. She (or he) is taking on the particular set of meanings and associations vehicled by the anti-whaling discourse that made these forms of consumption possible in the first place.

This is more than conforming one's behavior to the anti-whaling norm, and it may well be a sign of its successful internationalization. However, what the chapter brought into perspective that eludes the focus on norms and socialization is the way in which meaning is implicated in the process. It showed, first, that to internalize a norm is in fact to take on a particular set of meanings. Second, what actually drives the social actor to take them on is that these meanings serve to *identify* it, in the dual sense that they mark the actor in a particular way vis-à-vis social others and that it actively identifies with them, it recognizes itself in them. Hence the study of norms is indissociable from the study of identity, and appraising the effectiveness of a norm requires unpacking the way in which it is mobilized in the making of a social actor's identity.

The Performativity of Identity

Chapter 10 shifted back to the other end of the levels-of-analysis scale and revealed the ways in which this individual-level anti-whaling identity mattered for states. Simply put, the chapter observed what a state is doing when it is stepping into the subject-position carved out by the anti-whaling discourse, that is, how it is crafting its identity. What the discourse perspective draws out is that "doing" is not just about maximizing (preconstituted) interests; it is also about the making of one's self. Because of the ways in which this particular discourse connects to broader hegemonic discourses regulating the society of states, in stepping into the anti-whaling subject-position a state is enacting particular articulations of its identity. First, it is performing itself as a good, green, and democratic state that listens to what "the people" want for whales. Moreover, it is also behaving as the good, neoliberal state, capable of governing over highly productive economies without having to intervene in them.

The chapter showed that what appears on one level as a norm of good behavior with regard to whales is revealed on another as forming part of the broader governmentality arsenal currently being developed by states in neoliberal regimes, which serves to substitute a direct, heavy hand in the economy (on the sovereignty model) with indirect ways of steering the economy toward maximum productivity (governmentality). These are, in fact, just as interventionist; only their modalities of intervention are discursive. In the context of the growing economic importance of the tourist sector, the anti-whaling discourse offers a way of promoting whale-watching by meddling with what people can eat or what they do with whales in other states. Indeed, the anti-whaling discourse construes whale-watching not just as the appropriate way of relating to whales but as inherently incompatible with whaling—when in practice they coexist. Importantly, this observation undermines the distinction underpinning the constructivist socialization literature between the logic of consequence and the logic of appropriateness (see, for example, Checkel 2005) by showing that states adopt the anti-whaling discourse *because it achieves both*. Nor can this be simply reduced to a failure of the internalization process. That is, by this account, they would be observing the norm primarily because of the instrumental payoffs attached to it (in terms of green votes, for example) and without really "believing" in it such that, in fact, they do not really care about whales.

"Subject-Positions" and the Study of Identity

That states may not care much about whales may well be true, but my broader point is that one remains stuck within this dichotomy (consequence/appropriateness) only when one assumes social actors with preconstituted subjectivities, instead of looking at the constitution of their *subject-positions*, which is what I have proposed in this book. The fundamental move that enabled this shift in the study of state identities was to draw the distinction between "subject-positions" and "subjectivity," which comprises an individual-level identity concept. The concept of subject-positions offers a way of analyzing state identities unencumbered by a host of issues unavoidably bundled up with the concept of subjectivity (such as questions of affects, of degrees of belief, etc.) and that remain effectively irresolvable at the level of the state. Ultimately, it spares the analysis from having to ask how much a state has internalized the norm in order for it to be "genuinely" recognizing itself in its prescriptions for good behavior, which, in turn, requires finding a way of somehow prying open the state's "heart" or its "head" in order to measure degrees of genuineness and how much they have to do with their "true" essence, or who it "really" is.

The discursive approach developed in this book de-essentializes the study of identity. It shows that states' identities are in their doing; they are not lodged between the ears, to string out the metaphor of the body politic. A state is what it says it is and how it performs itself in its relations with other states. Moreover, this disjoining of identity from essences opens up another distinction, between *national* and *state* identity, as "thick" and "thin" modalities of a state's identity. Without this modulation, it is hard to conceive how the very same state, such as the United Kingdom or Australia, switched so suddenly from a whaling to an anti-whaling positioning. A focus on degrees of internalization of the norm would lead to having to evaluate which of the two was the more genuine position, the one that was closer to who the state really was; and subsequently to excluding one of the two either as disingenuous or, alternatively, as not involving the concept of identity. Instead, I showed that both constitute two different performances of the same state's (two different) identities, thereby illustrating the inherently fluid and performative nature of state identities. My point here is not to excuse hypocrisy on behalf of states. Rather, I am simply explaining how it is possible for a state to take multiple stances, or suddenly switch between stances.

The final chapter of this book focused on the subject-position carved out by a new pro-whaling discourse in resistance to the hegemonic anti-whaling discourse. It examined the different types of social actors that have effectively stepped into this subject-position, from states to local communities. Again, this illustrates how the focus on subject-positions enables the capturing of political practices at all of these levels, because of the parsimony of its criteria for political agency, namely, being a social-speaking actor. Chapter 11 first analyzed the way in which the pro-whaling discourse was mobilized by certain states in the constitution of their national (rather than simply state) whaling identities, in defiance of the regnant anti-whaling norm, and sometimes regardless of whether they actually whaled. It then examined how the nascent pro-whaling discourse tapped into articulations of sovereignty, around the signifiers of "food security" and "sustainable use." This served to tease out a supranational level of resistance, where pro-whaling states were allying themselves with developing states that did not whale. Resistance at the substate level was drawn out by focusing on ways in which the pro-whaling discourse was being developed as a marker of cultural identity by local communities against their own anti-whaling states. The dynamics of resistance driving the pro-whaling discourse reveals unusual, transnational alliances between substate indigenous communities and foreign pro-whaling states, thereby undermining the validity of distinguishing too sharply between state and nonstate actors. Focusing on subject-positions moves the analysis away from such distinctions. "Cultural diversity" was the signifier linking together these different types of political actors and these different levels of resistance. In crafting this discourse of resistance, what all these actors are reclaiming is the power to set the terms with which both the norms by which they interact and indeed their own identities are defined.

The chapter illustrated how identities are modified by shifting power relations between the discourses through which they are reproduced. Thus while the whaling identity is indeed an old subject position, as part I highlighted, the subject-position currently carved out by the pro-whaling discourse is a new one. Rather than a simple whaling identity, the pro-whaling discourse performs instead an anti-anti-whaling identity. Consequently, the pro-whalers now *need* the anti-whalers just as much as the anti-whalers need the pro-whalers.

In conclusion, identity matters to international relations, and focusing on subject-positions offers a useful way of studying identity, because it takes into consideration all the main points raised in this chapter. It shows how discourses and agents coproduce each other, why identity matters in the study of norms, why analyzing the dynamic constitution of identity matters in the appraisal of interests, why "doing" is more than just maximizing one's interests, why the making of identity involves wrenching a "self" from an "other," how identity can change, and how the "other" thus excluded seeks to reclaim the power to define its "self." To the extent that this book will have opened up a space for critically examining the discourses we unthinkingly dwell in in the course of our everyday lives, or to the extent that it will have encouraged a more thorough examination of any one of these seven points in the study of international politics, it will have served its purpose.

Appendix
Whales and Whaling—A Table of Events

IWC and Whaling-Related Events		Relevant World Events
1904	Pelagic (Antarctic) whaling begins	
1914–1919		World War I League of Nations is founded
1925	Factory whaling begins	
1929	Decline of whale species (rights, blues) begins Whaling Act (Norway)	
1931	Geneva Convention for the Regulation of Whaling	
1932	Industrial agreement on whaling limits BWU is adopted	
1935–1939	Intensification of Japanese and German whaling	
1936	Bilateral Cartel agreement to limit whaling (U.K.–Norway)	
1937	First attempt at a multilateral agreement	
1938–1939	Informal agreement on overall quota	
1939–1945		World War II
1946	International Convention for the Regulation of Whaling (Washington DC) 12 member states	United Nations and FAO are founded
1948		IUCN is founded

IWC and Whaling-Related Events		Relevant World Events
1949	First Plenary Session of the IWC	
1951	Establishment of the Scientific Committee	
1959	Norway and the Netherlands resign from IWC	Antarctic Treaty
1960	Appointment of the "Committee of Three"	
1961		WWF is founded
1962	Norway and the Netherlands rejoin	UN General Assembly resolution on Permanent Sovereignty over natural resources Rachel Carson's *Silent Spring*
1963	The Quota Agreement Japan and the USSR become first and second whaling nations	*Flipper* (film)
1964	"Committee of Four" End of Dutch whaling End of British whaling	Moby Doll, first aquarium killer whale
1965	Hunt of blue whales and humpbacks peters out	
1966	End of New Zealand whaling	Animal Welfare Act (U.S.)
1967		*Common heritage of mankind* (UN General Assembly resolution)
1969		Endangered Species Conservation Act (U.S.)
1968	New Zealand resigns End of U.S. whaling	
1969		FoE and IFAW are founded
1970	Observers admitted into IWC	*Songs of the Humpback Whales* (music)
1971		Pelly amendment to Fishermen's Protective Act (U.S., sanctions at imports) Ramsar Convention (Wetlands)

IWC and Whaling-Related Events		Relevant World Events
1972	International Observer Scheme adopted at the IWC U.K. bans import of whale meat for pet food European Union ban on whale products	Marine Mammal Protection Act (U.S.) UN Conference on the Human Environment (Stockholm) UNEP *Limits to Growth* and *Blueprint for Survival* World Heritage Convention
1973		Convention on International Trade in Endangered Species (Washington)
1974	The NMP is adopted, and the BWU abandoned	Greenpeace is founded
1975		Bergen FAO conference on "Mammals of the Sea"
1976	IWC's Permanent Secretariat (Ray Gambell) is founded (Cambridge, U.K.) New Zealand rejoins	
1977	Restructuring of SC	UNCLOS III proceedings
1978	End of Australian whaling The IWC proclaims the International Decade of Cetacean Research	Marine Mammal Protection Act (New Zealand)
1979	IWC opens to the Press Indian Ocean Sanctuary established End of Antarctic whaling	Convention on Migratory Species Packwood–Magnuson Amendment (U.S.) Moon Treaty
1980	IWC bans factory ships (except for minke whales) Whale Protection Act (Australia)	*World Conservation Strategy* IUCN/WWF/UNEP *A Programme for Survival* (Brandt report)
1979 1982	21 new member states	Convention on the Conservation of Antarctic Marine Living Resources
1981	Canada withdraws from the IWC	

IWC and Whaling-Related Events		Relevant World Events
1982	Ten-year blanket moratorium adopted by the Commission "Commercial" versus "subsistence" whaling established Reservations by Japan, Peru, Norway, Iceland, and the USSR	UNCLOS signed World Charter for Nature (UN General Assembly resolution)
1983		Great whales placed on CITES Appendix I
1985	Moratorium comes into force NMP abandoned	Vienna Convention (ozone layer)
1986		*Star Trek IV: The Return Home* (film)
1987	Japan withdraws its moratorium objection End of "commercial" whaling in Japan	*Our Common Future* (Bruntland Report) Montreal Protocol to the Vienna Convention Precautionary principle first emerges in International Law (North Sea Conferences)
1990	End of Soviet whaling End of Icelandic whaling	
1991	The Commission endorses stock-specific estimates of the SC	*Caring for the Earth* IUCN/UNEP/WWF HNA is founded
1992	The RMP rejected by the Commission SC chairman Philip Hammond resigns Iceland withdraws from the IWC NAMMCO established in the North Atlantic and adopts the RMP	UN Conference on Environment and Development (Rio) Climate Change Convention Convention on Biodiversity
1993	Norway resumes "commercial" whaling in its EEZ, using the RMP	*Free Willy* (film)
1994	Southern Ocean Sanctuary	UNCLOS comes into force

IWC and Whaling-Related Events		Relevant World Events
1995	The Commission endorses the RMP; the Revised Management Scheme remains under discussion	End Uruguay Round of GATT negotiations FAO Kyoto Declaration on Food Security FAO *Code of Conduct for Responsible Fisheries*
1997	Irish Proposal	WCW is founded Kyoto Protocol (Climate Change)
2001	Iceland rejoins	UNESCO Declaration on Cultural Diversity
2002		UN Earth Summit (Johannesburg)
2003		*Whale Rider* (film)
2006	Iceland briefly resumes whaling	
2007		UN Declaration on the Rights of Indegenous People
2008	78 member states	

Note: BWU, Blue Whale Unit; EEZ, Exclusive Economic Zone; FAO, Food and Agriculture Organization; FoE, Friends of the Earth; GATT, General Agreement on Tariffs and Trade; HNA, High North Alliance; IFAW, International Fund for Animal Welfare; IUCN, International Union for the Conservation of Nature; IUCN, International Union for the Conservation of Nature; IWC, International Whaling Commission; NAMMCO, North Atlantic Marine Mammal Cooperative Organization; NMP, New Management Procedure; RMP, Revised Management Procedure; SC, Scientific Committee of the IWC; UN, United Nations; UNCLOS, United Nations Convention on the Law of the Sea; UNEP, United Nations Environment Programme; WCW, World Council of Whalers; WWF, World Wildlife Fund.

Notes

Chapter 1

1. After all, we seem able to overexploit many other natural resources—tuna, cod, or sharks spring to mind—without "caring" so much about the animals in question.

2. For example, a second edition of Steve Lukes's 1974 *Power: A Radical View* was published in 2005. See notably the Review Symposium on this work, a recent edition of *Political Studies Review*, (2006, volume 4, number 2). The theme of the Journal *Millennium's* 2005 (volume 33, number 3) special issue was "Facets of Power," now turned into a book, *Power in World Politics*, edited by Felix Berenskoetter (2008).

3. That power circulates does not, however, mean that it is not located in the social structures, as I shall endeavor to show below. Hence I take issue with Barnett and Duvall's (2005) typology, which distinguishes too sharply between "structural" and "productive" powers.

4. Moreover, the debate about essences is also invariably linked to the even longer debate that has yet to take place within the social sciences about causality. Suffice it to say that a key milestone in Essentialist thought was posed by Aristotle, who shifted the enquiry away from Plato's focus on Ideas onto individual "beings" (*Ousia*, also sometimes translated as "substance" or "essence"). Aristotle's key move was to separate out, among the (four) causes of its coming into existence, the *ideational* from the *material* cause (see, for example, Irwin and Fine 1996).

5. The word "*sens*" in French means simultaneously "meaning" and "direction."

6. To be exact, the real object and the signifier are not one and the same thing in Saussurian linguistics. Both signifier and signified are linguistic phenomena (or else we would be back to a correspondence theory of the world, where the signified would be inherently linked to the object). "Signification" refers to the coming together of *signifier* and a *signified* to form a *sign*. Hence language is a *system of*

signs or more specifically in the Saussurian perspective of *signifying differences* (see Chandler 2007, Williams 1999).

7. A system of signifying differences is Saussure's definition of language—hence also, to paraphrase his famous thesis, in language there are only differences without positive terms.

8. This is not to deny constructivism's "critical ethos" (Price and Reus-Smit 1998, 285). By definition, taking a step out of the social world in order to examine how it was constituted is a critical step driven by normative concerns. However, it is also the case that *practical* power relations are seldom the starting point for this endeavor in this literature.

9. Thus Pierre Bourdieu's notion of "symbolic domination" encompasses, and operationalizes for discourse, without reducing it to a priori class interests, the Gramscian notion of "hegemony, as a combination of coercion and consent (Gramsci 1971, Forgacs 1988) that was ushered into the discipline by "critical international relations theory" in the wake of Robert Cox (1993), Stephen Gill (1993), Leslie Sklair (2000, 1995, 1994), and Randolph Persaud (2001), among others.

10. As we shall see below, despite a quasi-ideal-type speech situation (equal formal voting powers), it is less than clear that serious deliberative arguments take place at the IWC.

11. My point here is not to separate out once again the social from the discursive after having attempted to demonstrate how they mutually constitute each other but simply to show that, while the literature on ideas *does* account for their social or institutional contexts, they tend to forget about discourses, precisely because they overlook the discursive nature of the social.

12. For a theoretical overview of these two lineages, see notably Williams (1999).

13. This is the key difference with a Marxian critique of ideology, ultimately geared toward uncovering the underlying truth obscured by the "false beliefs" nurtured with ideologies. This perspective, by contrast, does not presuppose some preexisting, universal Truth. It seeks rather to gauge the political *effects* of claiming that something is true.

14. In view of the evacuation of power that has accompanied the use of the term in the study of international politics, to the word "norms" I prefer Pierre Bourdieu's term, *nomos*. The *nomos* similarly refers to the normative order located within the ambient discourses. Much stronger than the now much diluted term "norms," *nomos* refers to the unexamined matrix of all thought, speech, and action, as well as their legitimizing principle (Bourdieu 2003).

15. This is another way in which the discursive approach deployed in this book differs from the Marxian ideology critique (see, by contrast, Hawkes 2003, Lee 1992, Thompson 1990). It is not driven by any teleological aim to bring the "truth" to social actors shrouded in false consciousness, because it harbors no essentialist presupposition as to how the actors actually are (as to their essence, in other words).

Chapter 2

1. I wield the categories of "East" and "West" here *à la* Max Weber as a descriptive, topographic shorthand; but I am well aware of the essentializing, orientalist underpinnings for which they have been amply critiqued. Yet these categories are hard to avoid not only because they are widespread in both the literature and the discourses on whaling but because the anti-whaling discourse—the main object of analysis—is itself, I argue, built on similarly orientalizing moves. Thus after "speaking with" them, I eventually draw these moves out in the latter part of the book (chapters 8–11).

2. Baleens serve to trap and filter food organisms. Of the six families of whales, only three, known together as the Mysticcti (or "mustachio whales") possess them: the right whales, rorquals, and gray whales. The sperm, beaked, narwhal, and beluga whales are toothed carnivores. All, however, have oil (Ellis 1992).

3. Chemically, whale oil is simply fat, whereas sperm oil, extracted from the head of sperm whales, is unique in composition.

4. An abridged version of the 1970 two-volume monograph on the modern whaling industry written by the Norwegian historians Johan Nicolay Tonnessen and Arne Odd Johnsen at the behest of the Norwegian whaling industry and the Norwegian government was translated into in English in 1982. It remains to this day the most comprehensive historical analysis of the industry around the world and is thus a key source for the historical chapters of this book (chapters 2–4).

5. The West's loss of control over whaling is a key factor in explaining the emergence of international whaling regulations, as we shall see in chapter 4.

6. As a lubricant, sperm whale oil is superior because it remains as a surface film, while other oils (including baleen oils) drain away from lubricated surfaces. Furthermore, it is more long lasting than synthetic oil (it does not dry out). It was allegedly used in the making of the Little Boy, the first atomic bomb.

7. This was actually a return to whaling, for the Germans had been whaling around the coast of Greenland in the eighteenth century (Tonnessen and Johnsen 1982).

Chapter 3

1. Although this CIA document constitutes in itself a discourse, in this particular instance, rather than taking the critical step out of the discourse, I step *into* it, for the purpose of showing how states still talked about whales and whaling in the mid-1950s. Thus this discourse is treated as a straightforward historical recording of states' perceptions of whales and whaling under the whaling order rather than as material to be subjected to a critical discourse analysis, as undertaken in chapter 8.

2. Other national cetological research programs included the Dutch Werkgroup Walvisonderzoek (set up in 1947), the Labortoire de Biologie des Cétacés et

Autres Mamifères Marins (Ecole des Hautes Etudes, established in 1950), the humpback research programs developed in New Zealand and Australia in 1949 and 1951, respectively, and finally the U.S. Fish and Wildlife department, which developed a research program in connection with Californian whaling in 1956 (Matthews 1975, 172; Small 1971).

Chapter 4

1. It is now a commonplace in political science that states' interests are not given but socially constructed. What I present here are the mechanisms through which this occurs.

2. For a classic, constructivist analysis relationship between state interests and identities, see notably Wendt (1999). Here, however, I am departing from the IR constructivist approach and stepping into a more strictly sociological perspective.

3. While Bourdieu granted a central place to language at large in the reproduction of social relations of power (see Bourdieu 2001), discourse was admittedly not his methodological starting point. Because, I argue, it need not be: the social systems he analyzed were mostly contained within the nation-state; thus discourses shared in common by the different social actors were a given rather than a constitutive feature of those systems. Methodologically, discourse simply was not the most pertinent unit of analysis for him, compared to other sociological factors. Here, discourse is what holds the entire social system together; hence it is at the center of the analysis. For these reasons, I use the terms "discursive order" and "nomos" interchangeably.

4. The reasons for my preference for the term "*nomos*" rather than "norms" are further expounded in the following chapter in the context of a broader engagement with constructivism and the literature on socialization in international relations.

5. On this point, it is interesting to measure the extent to which Japanese whaling is misrepresented in the literature on whaling. George Small (1971), in electing the two countries where whaling had the most and least regulation, selected Norway and Japan, ignoring the historical fact that whaling had remained completely unregulated in the most important whaling country (the United States) throughout the nineteenth century. Japanese whaling, on the other hand, far from being ungoverned, was enmeshed in the *Zaibatsu* system of formal and informal of relations that organizes the relationship between state and business in Japan.

6. Another way to understand this is by drawing on a distinction, which is more fully developed in chapter 10 where it becomes more central, between the *state* and the *national* identity. Thus the United States did actually continue to remain interested in being able to whale, and provided for this possibility within the IWC (see chapter 2), even though U.S. monies had largely pulled out of the trade,

and the state could therefore not be defined in terms of actual, material interests in whaling.

7. While the notion of a British whaling identity may seem awkward today, this, I argue, is precisely an effect of the anti-whaling discourse, whose key mechanism is amnesia, or the active denial of the West's whaling past (see chapter 8). However, the interests of a social actor define its identity insofar as what you are interested in defines who you are (see also Wendt 1999). What I capture here is the sharpening of the British state's interest in whaling and, therefore, its whaling identity. It is also worth remembering that the United Kingdom caught, by far, the largest number of whales in total over the twentieth century (Schneider and Pearce 2004, 546).

8. The right whales have historically been the most hunted whales, because they are the easiest to catch: they are slow swimmers and are the only whales whose carcass does not sink, which makes them the "right" whales for whalers (the actual origin of their name, according to Ellis 1992).

9. There are, however, many problems with the industry's catch data, whose reporting to the Bureau depended on the goodwill of the nations. For example, it was discovered in the 1990s that Soviet catches had been systematically misreported for decades. These problems, however, while they generally affect the calculations of whaling statisticians, make no difference in the broad catch trends sketched out here.

10. The twelve original member states included the USSR, the United Kingdom, the United States, France, Denmark, Australia, Norway, Argentina, South Africa, Japan, Iceland, and Mexico.

11. Article II (4) of the ICRW establishes a "Contracting Government" as "any Government which has deposited an instrument of ratification of his given notice of adherence to this Convention." In practice, those countries that have not renewed their annual membership fees are not considered members by the IWC Secretariat at that year's annual meeting.

12. This directly undermines M. J. Peterson's assessment that the United States refrained from using its power in the IWC. Stating, as he does (1992, 149), that "hegemony" was simply not at work in the IWC and that the United States did not use its "power" before 1970 is possible only by ignoring discursive understandings of both "hegemony" and "power."

13. Article X (1), (2), and (3) stipulate that all instruments of ratification are to be deposited with the government of the United States, who is to inform other signatory governments of all new accession instruments received. Interestingly, this provision was not altered with the creation of a permanent ICW Secreteriat in Cambridge, United Kingdom in 1975.

14. The legal confusion was cleared in an extraordinary meeting of the IWC, and Iceland's membership rights were fully reinstated in 2002.

15. In 2003 the IWC adopted a resolution on simultaneous translation, whose costs are born entirely by France, which was the only country to volunteer additional funds to this end.

Chapter 5

1. Partly as a result of its becoming a dominant global discourse, the anti-whaling story line is dispersed across a wide variety of media, increasingly visual as well as written, as we shall see in this chapter. The purpose here is simply to draw this story-line in broad brushstrokes as the backbone of the anti-whaling discourse. The discourse itself is analyzed more closely in chapter 8. Equally as a result of this dominance, this story-line runs through activist (Hunter 1980, FoE 1978, Day 1992) as well as academic accounts of the whaling issue (Scarff 1977, Small 1971, Birnie 1985, Peterson 1992, Caron 1995, M'Gonigle 1980, D'Amato and Chopra 1991, Simmonds and Hutchinson 1996, Mitchell 1998), both of which constitute sources here.

2. This wave of popular literature was triggered in 1948 with the publication of both William Vogt's *Road to Survival* and Fairfield Osborn's *Our Plundered Planet.* Then came Rachel Carson's *Silent Spring* in 1962 (thirty-one weeks on the *New York Times* best-seller list), and the wave accelerated toward the late 1960s, with Paul Ehrlich's 1968 *Population Bomb*, Barry Commoner's 1971 *Closing Circle*, and the string of publications around the Stockholm conference, such as the Club of Rome's 1972 *Limit to Growth*, the *Blueprint for Survival* published by the team of *The Ecologist*, Ward and Dubos's 1972 *Only One Earth*, and Schumacher's 1973 *Small is Beautiful.*

3. The term "floating signifier" was first introduced by anthropologist Claude Lévi-Strauss to refer to signifiers or words that operate with no fixed signified or referent. It has been applied to the analysis of concepts such as "race" or "gender," but see also Rubenstein (1989) for a use of the term to analyze Greenpeace practices.

4. A "synecdoche" is a rhetorical figure whereby referring to the part (the sail) conjures up the whole (a boat). It is considered here as a special form of "metonymy" (Chandler 2007).

5. This is not to say that it was the first instance of interstate environmental collaboration—in fact, such collaboration was in place since the early twentieth century (see Epstein 2006). However, it occurred mainly as a series of disparate initiatives on specific issues (such as birds, seals, national parks, etc.). The Stockholm conference was, however, the landmark conference that established "the environment" as such on the agenda of international cooperation (Thomas 1994, Vogler and Imber 1996). Attended by 113 states, it was also the first truly global, rather than regional, environmental convention.

6. In February 1966 the popular magazine *Life* published a very successful issue on "Concentration Camps for Dogs" (volume 60, number 5). The 1960s also witnessed a wave of new animal protective societies that were instrumental in the obtaining the passage of these laws. Their lobbying strategies are discussed in the following chapter.

7. The Delmonte company was still able to catch whales after this first ban (notably the most endangered gray whale) by using the "scientific permits" provision

of the ICRW (article viii)—the same legal loophole used by Japanese scientific whaling today. Thus, in addition to this law, Nixon's 1971 executive order was necessary to do away with American (nonaboriginal) whaling altogether.

8. The endangered species list system was taken over from the IUCN, which had been listing endangered species in its "Red Data Books" since 1963; however, this was the first time it was translated into policy.

Chapter 6

1. See specifically Article V.2.b. of the ICRW, which establishes that any "[amendment to the Schedule] shall be based on scientific findings."

2. Although Peter Haas (1992, 23) states that his is a "consensus theory of a finite and temporally bounded notion of truth" rather than a "correspondence theory of the world," and that his concern is with "consensual knowledge" regardless of its truth content, in effect his approach rapidly slips back toward a correspondence theory of the world. For he starts by positing that consensual knowledge about the world is possible (Haas 1992, 23). However, because he underplays the role of discursive power, not much is left to drive this consensus aside from some implicit notion of truth that is objectively attainable through the proper exercise of reason, effectively the cornerstone of a correspondence theory of the world. This is confirmed by his development of the ways in which scientists influence policy makers. Specifically, scientists "elucidate cause-and-effects relationships" (Haas 1992, 15). Thus unlike in Maarten Hajer's (1995) analysis, where the cause–effect links drawn by experts are analyzed as part of merely one (slightly more powerful) possible story line, and where the question of objective truth is therefore successfully suspended, in Haas there is an underlying sense that if policy makers turn to scientists to elucidate these relationships, it is because the latter really can do so, albeit with the right conditions (see also Haas 2004). Altogether the power of science thus tends to escape critical scrutiny.

3. This periodization builds on, and extends, John Gulland's (1988). See also Mitchell (1998).

4. By contrast, from 1977 onwards, a full time editor was employed and the *Report of the International Whaling Commission* was professionally typeset; and by the early 1980s a review process was established (Donovan 1999).

5. However, the very first transnational scientific unions were actually set up earlier: the International Union of Biological Science, for one, was established in 1919, and the International Council of Scientific Unions in 1931.

6. This culminated with Raymond Beverton and Sidney Holt's 1957 *On the Dynamics of Exploited Fish Populations.*

7. The overall quota set by the 1946 Convention was 16,000 BWU, but the whaling industry maintained discretion over which species were caught. Since 1 blue whale was equivalent to 2 fin whales, 2.5 humpback whales, or 6 sei whales, there was an incentive to catch the second largest species once the blues had disappeared.

8. One could object at this stage that the previous phase was simply a period of bad science, and people began paying attention once the science became good. However, what this story shows is that the science continued to improve and was increasingly ignored.

9. The main idea behind MSY, which was first developed in fisheries management, is that, following the laws of population dynamics, reproduction rates in a given population tend to increase in reaction to the rise in death rates, whether due to harvesting or natural causes. The MSY is the point beyond which a population tends to decline overall, once the negative effects of the increased death rates start to exceed the positive effects of increased reproduction rates. So long as catches are kept below the MSY, the "surplus population" can allow for sustainable, long-term harvest. The MSY level for whales has been estimated at 60 percent of initial stock size. The NMP classified all whale stocks into three categories: (1) initial management stocks, which numbered at least 72 percent of original stock size; for these catch limits would be set always below the MSY (at 90 percent of MSY); (2) sustained management stocks, those between 54 percent and 72 percent of the original stock size; catches were limited to 90 percent of MSY or less; and (3) protection stocks, below 54 percent of original stock size, for which no catches would be allowed.

10. Discursively, it is significant that the categories used for this nomenclature are those of the IUCN's Red Data book, taken over, in turn, from the U.S. endangered species list. Yet these were initially developed to list only *endangered* species. Listing all marine mammals, including many that are not endangered (as is explicit in the volume itself) in this way nonetheless invokes, by association, a generic sense of endangeredness.

11. At the preparatory meetings to the Bergen conferences (in Boston) Fortom-Gouin's presentation had been preceded by a presentation by the neuroanatomist George Pilleri, who had specialized on whales and dolphins for over a decade, extensively demonstrating the absence of intelligence in marine mammals (interview with Ray Gambell). For one of the more sophisticated articulations of a preservationist argument premised on cetacean intelligence, see D'Amato and Chopra (1991). For a critique of the lack of scientific basis to the claim of whale intelligence, see the analysis of cetologist Margaret Klinowska (1994).

12. As I have argued extensively elsewhere (Epstein 2005), a survey of thirty-five years of the *Reports of the Scientific Committee* reveals that SC's research program tended to mirror the concerns of the time. The issue of humane killing, for example (the study of killing methods in order to minimize the animal's suffering), was not a new concern the SC, yet it had fallen to the wayside by the late 1960s, eclipsed by the issue of quotas and no doubt also because of a lack of interest on behalf of the Commissioners. In 1975, as the anti-whaling discourse was becoming increasingly entrenched, the topic was successfully reestablished on the research agenda by the American Commissioner and has since remained a standard research item. Here are some more recent examples of how discourses circulate from one research setting to another: in 1993 the SC added the effects of

global warming and ozone depletion to its research agenda. Then in the wake of the mad cow crisis both in Europe and in Japan, the IWC has been passing resolutions on the contamination of whale meat and the threat to human health since 1998.

13. While this may sound like a typical epistemic community story, it is not, in that it was an attempt to generate consensus on a topic that has not yet been adjudicated on. In other words, the science was not yet in on the issue.

14. They "urged governments and intergovernmental organisations to give support to the major campaign for conservation of marine species ... launched by IUCN and WWF, and, as opportunities arise, that governments facilitate the implementation of such projects" (FAO 1978, vol. I, 32).

15. "Cetaceans in captivity" is explicitly mentioned in the Bergen conference Report (1978, xx) as one of the "important scientific advances" that contributed to this global consensus regarding cetaceans.

16. One example of a cultural event that was also a "pop science" celebration of whales was the 1974 University of Indiana (Bloomington) conference, a "landmark event," in Sidney Holt's own words (Holt 2000) that brought together "hundreds of scientists, economists, sociologists, entrepreneurs, engineers ... musicians, poets and other writers," in a "celebration on a grand scale."

17. de la Mare ran a series of simulations that demonstrated the inadequacy of the process from the incorporation of the data to the yielding of a management quota, which presumed a flawless (and thus unrealistic) data set (interview with Justin Cooke).

18. Part of the difficulty is that, not only are these whale population estimates extremely difficult to undertake, but given the high rates of uncertainty involved, they are often contentious. Moreover, they often need to be re-started once completed to keep track of often rapidly changing populations. Thus the 1990 figure for minke whales is contested and at this date under review. The latest update as to what populations for which reviews have officially been completed or are still under way can be sought on the IWC website.

19. Cooke's procedure was also the most generic and thus had the added benefit of simplicity, even if it meant, in Philip Hammond's words, that it "must be unnecessarily conservative in some cases" (personal correspondence with Philip Hammond).

20. Both Norway and Iceland had lodged reservations to the whaling moratorium.

Chapter 7

1. Among the other activists heard by the members of congress that same day were the FoE "wildlife consultant" on the intelligence of whales, as well as the president of the Humane Society U.S., Brian Davis of the IFAW, and Miss

Harris, daughter of one of the sponsors of the Harris–Pryer Bill (U.S. Congressional Record 1971).

2. For example, although it was never actually certified, Iceland constantly revised its scientific research programs between 1986 and 1988 in response to threats to third countries. When Japan initially lodged a reservation to the moratorium in 1982, it was not in fact certified; yet the mere menace generated important domestic divisions that had to be resolved in the High Court, and Japan eventually retracted its reservation in 1987 (DeSombre 2000, 209).

3. However, this "native exemption" has not been equally available to all U.S. native whaling people. For example the Makah in the State of Washington were only able to resume their traditional whale hunt of gray whales in 1999, once the species had been officially removed from the Endangered Species list in Washington D.C. This, despite the 1855 Treaty of Neah Bay, which ceded Makah ancestral land to the U.S. government against a guarantee of protection of the tribe's "right of taking fish and of whaling or sealing at usual and accustomed grounds and stations" (Article 4). The Makah simply lacked the leverage afforded to the Eskimo by oil-rich lands (see also chapter 11).

4. The issue, however, had caused a rift within American NGOs themselves. The American branch of FoE, for instance, was in favor of filing the objection, which was opposed by most other groups and other branches of FoE.

5. Thus my only contention with M. J. Peterson on this point is that she sees this coming into effect only in the late 1970s, when it was at play since the creation of the IWC (see chapter 4).

6. This created considerable confusion. The Panamanian government claimed it had not been consulted. Meanwhile *Eco* reported a visit of the Japanese trade delegation to Panama and accused the Japanese government of applying economic pressure by threatening to cancel a U.S. $10 million sugar purchase, which was emphatically denied by the Japanese (Day 1992).

7. These proposals have so far been unable to muster the three-quarter vote required for schedule amendments, which sanctuaries (just like moratoria) are considered to be.

8. The connection used was a villa in the Seychelles owned by the Pahlavi family, who also funded the London-based Threshold foundation for which Watson was acting Secretary General.

9. For example, the IWC's 55th annual meeting in Berlin in 2003 was opened by an anti-whaling address by the environment minister Renate Kunast, which had been drafted with the help of Greenpeace Germany (interview with Thilo Maack).

10. Thus Kenya's well publicized efforts to ban the trade in elephant ivory in 1989 also have to be seen against this backdrop: in 1988 the United States had passed the African Elephant Conservation Act, which created a fund for disbursing aid to African countries committed to its preservation, of which Kenya became a major recipient (see Epstein 2006).

11. This person was closely involved with the anti-whaling campaign throughout the 1980s and was looking back on the whole period. Given the sensitive nature of the material and the intense politicization of the issue, this person has requested anonymity.

12. This information was confirmed to me, albeit in less explicit terms, by the then IWC Secretary, who personally received the cheque.

13. In 1984, Palacio's credentials as Alternate Commissioner for St. Lucia were revealed as forged, and he was banned from the IWC meeting. St. Lucia has now become a pro-whaling voting country. Many of the credentials used by activists originated not from the country's capital but from diplomatic missions abroad.

14. Roger Payne regularly appeared on delegations for Antigua and Barbuda from 1983 to 1987, until his credentials were questioned by Japan, and he was dismissed from the IWC. In 1983 the Commissioner for the country was P. O. Spencer, also registered as Alternate Commissioner that year for St. Lucia, which effectively conferred upon one man alone the voting powers of *two* foreign countries. Lastly Fortom-Gouin would be Alternate Commissioner for Antigua and Barbuda from 1987 to 1989—in 1990 the country briefly resigned from the IWC. Today Antigua and Barbuda is once again an IWC member, as a supporter of sustainable whaling, and its Commissioner, Daven Joseph, is one of the most outspoken against the anti-whaling pressure (which he attributes equally to foreign governments and NGOs).

15. Holt would also become Alternate Commissioner for Seychelles, with voting powers, from 1984 to 1992. Seychelles would withdraw its membership in 1995, at which point Holt reappeared as an observer on the Italian delegation.

16. St. Vincent and the Grenadines, which was enrolled in the anti-whaling camp, is actually a whaling country (specifically the Island of Bequia). After the moratorium vote St. Vincent dismissed Davey and appointed a St. Vincent national (Gloria Peningsfeldt), who obtained the recognition of Bequian whaling under the aboriginal subsistence scheme, under which a small quota of humpback catches continues today.

17. The "moratorium" has since become a key instrument of global environmental activism/policy making. It was used again in the European Union "genetically modified organisms" campaign in the late 1990s.

Chapter 8

1. A predicate can be logically contained in the verb itself, such that the most elementary sentence structure is of the form subject–verb and the sentence can stand without an object or adjective after the verb.

2. This is why the discourse analysis developed here is quite removed from any concern with "false consciousness," nor is it a method for uncovering the "true" identity of individuals that would have been masked by their being "duped" by a

dominant ideology. To reiterate an important point, this suspension of the truth/falsity question distinguishes this study of a dominant discourse from the "ideology critique" approach.

3. Some anti-whaling boycotts were very effective indeed. The 1988–1990 Greenpeace boycott of Icelandic fish products was so effective (50 million U.S. dollars of loss in export revenue) as to bring Icelandic whaling to a halt. During the 1980s Iceland practiced some scientific whaling, the only legal form of whaling it could engage in, since it had retracted its initial reservation to the IWC commercial whaling moratorium in 1983. Interestingly, Iceland was never in fact sanctioned or even certified by the U.S. State Department for its scientific whaling (DeSombre 2000). This NGO boycott thus constituted one instance where whaling was brought to an end entirely by NGO, rather than state, pressure and despite a national opinion that remained strongly in favor of whaling (Brydon 1991, 1996; see also chapter 10).

4. This reflects the trajectory of my own research process: after five years (2000–2005) of collecting textual material, the question arose as to how best to use it. After initial attempts at coding all anti-whaling texts collected at two IWC meetings (2001 and 2003), I came to the realization that this systematic cross-text analysis yielded exactly the same categories as an in-depth analysis of one text carefully chosen according to historical and strategic considerations.

5. This has been the cause of increasing divisions among anti-whaling NGOs. For a group like WWF, for example, which has a wide and heterogeneous membership, it is becoming increasingly difficult to position itself on this question. At the IWC it continues to hold a no-whaling stance; yet some local branches have parted ways with it to support a "sustainable whaling"—for example, WWF-Japan (Brown 2002), or WWF-New Zealand, which has backed the Maori request for sustainable use of whaleparts (Te Ohu Kai Moana 1997).

6. More than just a mechanism of self-defense, this explicit positioning of the discourse itself vis-à-vis the label "racist" is not without evoking reflections on the emergence of new forms of "neoracism" in the late twentieth century, where cultural signifiers have replaced biology as the basis for segregationist practices of "othering," and which explicitly define themselves as not racist (Hardt and Negri 2001, 190–195; Balibar 1991).

7. Consider this abstract from *Moby-Dick* (1851), chapter 41: "Nor was it his unwonted magnitude, nor his remarkable hue, nor yet his deformed lower jaw, that so much invested the whale with natural terror, as that unexampled, intelligent malignity which, according to specific accounts, he has over and over again evinced in his assaults. More than all, his treacherous retreats struck more of dismay than perhaps aught else. For, when swimming before his exulting pursuers, with every apparent symptom of alarm, he has several times been known to turn round suddenly, and, bearing down upon them, either stave their boats to splinters, or drive them back in consternation to their ships."

8. I owe this point to anthropologist William Simmons (Brown University).

Chapter 9

1. "Nonconsumptive utilizations of the whale" is a discursive category created by NGOs themselves. Its origins are traceable to the 1983 NGO conference "Whales Alive" in the Seychelles. However, it rapidly spilled over into IWC debates and was taken on by anti-whaling states (it surfaces, for example, in the various "sanctuary proposals"). Yet ambiguity persists as to whether it technically falls under IWC competence or not. In 1984, for example, the IWC set up a "working group" to determine the status of the NGO recommendations report. In the end the Commission simply "took note of the report," without taking any actions nor resolving the issue of competence (IWC 36 1984, para. 16).

2. Article V.2 of the ICRW requires taking "into consideration the interests of consumers of whale products and the whaling industry."

3. For example, the dolphins of the Monkey Mia Western Australia, as a result of human contact, began developing new diseases and new patterns of dependency and even attacked the visitors. The program continues, but it is now forbidden to "touch" the dolphin (one can only hold out a fish to the animal, under the tight gaze of a ranger).

4. These rates were drawn from the Web site of the Norwegian company Whale Safari www.whalesafari.no/whalesafari/rates, accessed 03/03/2008.

5. In 2004, for example Greenpeace's income (Greenpeace 2004a) was comparable to the turnover of a small multinational company such as Moulinex.

6. My point here is not to suggest that Greenpeace as a whole is driven exclusively by instrumental, wealth-maximizing concerns. Greenpeace's campaigns are often very valid and key to drawing attention to issues that may otherwise go unnoticed as indeed the whaling case itself has shown. I do, however, think that, because of its special status in the NGO's history and recruitment strategies, the whaling issue has become largely instrumentalized.

7. For example, the week preceding the IWC 2001 annual meeting in London, photographs of slaughtered whales copyrighted for IFAW regularly featured in the British *Daily Mail.*

8. Indeed, meaning, as we recall, is not inherent to the terms of a language but born of their difference, generated by their interplay.

9. This is not to say that the whale does not genuinely mean something very specific and personal to the individual who purchased the whale object (based, for example, on some personal experience with whales, etc.). However, nor is this signification a completely individualistic one in an anti-whaling society, insofar as it emerges against a backdrop of preexisting significations projected onto the whale, and it serves to reproduce them.

10. This is not to deny that there are needs outside of the consumptive/productive system, nor is it to say that *all* needs are produced within the system, which would, once again, amount to evacuating human agency. It is rather to

consider the specific ways in which *some* needs may be generated or harnessed by the system.

11. Individual donations comprise 131 million euros of the 158-million-euro income (Greenpeace 2004a). Concretely, each person adhering to Greenpeace International (without transiting via a local branch), pays in an automatic monthly "donation" of between 10 to 30 dollars through the "Rainbow Warrior giving plan." Those who prefer can contribute a "single gift" of anything between 25 and 100 U.S. dollars (but no less than 5 dollars, Greenpeace 2004b).

12. What I am describing here is an *ideal-type* identity, of the sort that marketing companies are used to wielding, rather than a particular person.

Chapter 10

1. For a similar rejection of the analogy between persons and states, see notably Wight (2004) and Flockhart (2006). However, instead of turning away from psychology, which I see as the logical implication of such a rejection, Flockhart embraces its tools, albeit those of a social rather than an individual psychology (but the former also necessarily stems from the latter). Once again, this, for me, is an unnecessarily essentializing move (see chapter 1).

2. Indeed, the terms by which Croatia, a new member of the European Union, justified its accession to the IWC in 2007 to the BBC suggests that this marker of suitability is still very much at play (Black 2007).

3. This is not to say that environmental NGOs are unanimous on the whaling issue. For example, WWF Norway, WWF Japan and WWF Denmark have explicitly rejected the anti-whaling stance, in contradistinction to most other local WWFs (Christoffersen 1994, Brown 2002). Other NGOs, such as FoE, have distanced themselves from the issue altogether as it became increasingly polarized. However, Greenpeace Japan has adopted the anti-whaling stance.

4. Canada's positioning is interesting. Since resigning from the IWC in 1981—just before the commercial whaling moratorium went through—Canada has continued to whale. It is thus technically a whaling country; however, unlike Japan and Norway, Canada has been rather coy about its whaling identity at the international (but not domestic) level. Canada's quiet positioning thus illustrates a different, less outward-orientated modality of discursive identity construction.

5. This "playing well" is not just a product of rational calculation but something more implicit, more tacit, whereby one is also "playing into" the rules of the system (and thus also unknowingly reproducing it).

6. See chapter 4 for an introduction to Foucault's governmentality thesis. Here it is worth recalling that, while governmentality was coined as a historical concept (as it was used in chapter 4), it was also specifically developed to account for the new forms of neoliberal governance that were taking shape at the time of Foucault's writing (Foucault 1991, 2003).

7. To reiterate, the point I am making here is not a direct comparison between whale meat and beef; rather it requires a broader focus on the range of governmental strategies developed by the contemporary neoliberal state.

Chapter 11

1. For example, in his opening address at a conference on Whaling in the North Atlantic in Reykjavik in 1997, Minister for Foreign Affairs Halldór Ásgrímsson declared: "It is understandable that environmental campaigners should focus on endangered species, and it is also understandable that their arguments about whales should appeal to nations that have little acquaintance with fisheries."

2. This was legally possible because Norway, Japan, and Iceland all have reservations placed on the minke whale ban, thereby allowing trade in minke whale parts between these three countries. (Norway and Japan both placed their reservation in 1983; Iceland joined CITES in 2000 with a reservation on minke whales.)

3. For example, in the Bendiksen family in Lofoten, three out of four brothers are whalers, as well as the cousins on either side (interview with Bjorn-Hugo Bendiksen).

4. By contrast the blubber is consumed in Japan and the Faeroe Islands. The Inuit (notably in the Faeroe Islands) have traditionally considered the blubber a delicacy (*mattak*). Norway shipped its first export (30 tonnes of meat and blubber) to Iceland in 2002, and a second batch to the Faeroe Islands in 2003. However trade with Japan has stalled, as Japan continues to be hesitant on whether to take this route. Currently, the blubber is stored in cold rooms (interview with Rune Frovik, former HNA secretary, and with Jan-Odin Olavsen, whaling captain and proprietor of a whale processing plant, Lofoten Island, Norway, summer 2003).

5. For example, Lars Walloe, the main scientist on the Norwegian delegation, opened Norway's intervention on the floor of the IWC Plenary by emphasizing that "Norway has been heavily involved in both this [climate change] research and the scientific discussions as well as in the international political process from the very beginning, when these environmental problems were first recognized by the scientific community. For example, Norwegian scientists working on chemical reactions in the atmposphere were among the first to publish papers on the processes resulting in the breakdown of the ozone. Norway was also one of the first countries that was most active in the political process that resulted in the Montreal Protocol on the release of CFC [chlorofluorocarbon] gases" (Lars Walloe/Statement by Norway, 1999).

6. For example, in 2002, for the first time ever, Japan retracted its traditional support for the bowhead quota requested by fellow whalers, the American Inuit: since it was not getting any support for its Small Type Coastal Whaling proposal from the U.S. delegation, it would no longer bring its vote to an American

request. The turnabout caused surprise in the Commission, and havoc in one of its only smooth-running sections (the Committee on Aboriginal Subsistence Whaling). The issue had to be resolved at an intersessional meeting. In this way Japan brought the United States back to the negotiating table, ready for the first time to consider Japan's demands. After this episode Komatsu, who masterminded the move, was "promoted out" of his seat as Alternate Commissioner, allegedly under diplomatic pressure from the United States. However, he was back again at the next IWC meeting in Berlin in 2003.

7. Two weeks after the results of the government poll were published, the newspaper *Asahi Shimbun* conducted its own poll, which showed that only 47 percent of the public agreed with it (Brown 2002).

8. Furthermore, it is noteworthy that diplomatic pressures tend to apply on both sides. For example, at the 2003 meeting, prior to the vote on the South Pacific Sanctuary proposal tabled by Australia and New Zealand, the Commissioner for the Solomon Islands, Nelson Kile, was approached by the delegates from "the two superpowers in his region" and asked about his voting intentions on their Sanctuary proposal, to which he declined to answer. A few hours later Mr. Kile, who as fisheries minister had been given carte blanche from his prime minister back home, received a personal fax from his prime minister asking him to support the proposal. The Australian and New Zealand delegates then returned with copies of the same fax he had been sent (interview with Nelson Kile). The construction of the "Japanese vote-buying" narrative thus simply frames out the fact that, as this incident illustrates, interactions at the IWC are inseparable from a much broader interplay of North–South diplomatic pressures exerted around the disbursement of aid monies, in which many different issues are often bundled together. It is also interesting that this type of incident, which would undermine the credibility of their key state allies, should escape the scrutiny of anti-whaling NGOs.

9. The analysis of sustainable development as a governmental discourse runs roughly as follows. The concept of "sustainable development," under increasing criticism of late, is more than just an "oxymoron" that reconciles the irreconcilable (Redclift 2005) or a concept so vague as to be devoid of meaning (Sachs 2000, Connelly 2007). Or else why would it be so successful? Because it is a governmental discourse that serves to orchestrate the turn toward a neoliberal "green governmentality." Luke points to the host of power/knowledge formations ushered in around new types of "ecological expertise" and to the ways in which it has triggered the deployment of new political technologies for the management of resources and thus populations, both at home and in developing countries (see also Adams 2001 for the latter). See also Khuels (1998) in particular for the relationship between sovereignty and the environment exposed as the locus for the deployment of governmentality.

10. In 1996, the Makah were granted a quota of 20 Pacific gray whales, with the backing of the U.S. government, but without the support of the working group Chairman. They caught their first whale only in 1999, however, at which point the whalers were met with a jeering, angry crowd (Porterfield and Denn 1999,

Anderson 2001), notwithstanding the "strong comeback" of the Pacific gray whales (Barber 1999).

11. Today there are many pro-whaling NGOs, as evidenced by the list of delegates at any of the IWC meetings. Here I focus on the organizations that have been proactive in entering global whaling with the explicit purpose of reshaping people's mind-sets and, thus, sought to match the anti-whaling NGOs on their own terrain, rather than organizations that simply support whaling back home.

12. Since then, their application for subsequent whaling permits have been tied up in court disputes (Lewis and Shukovsky 2007).

13. See, for example, http://luna.pos.to/whale/.

14. In fact, the Maori's first attempt to alter New Zealand's anti-whaling policies originated in the debates preparing for New Zealand's negotiating stance in the Rio Convention (correspondence with Sean Kerins, June 2004).

Bibliography

Adams, W. M. (2001), *Green Development: Environment and Sustainability in the Third World*, London and New York: Routledge.

Adler, E. (2006), "Constructivism and International Relations," in *Handbook of International Relations*, Carslnaes, W., Risse, T., and Simmons, B. A. (eds.), London, Thousand Oaks, and New Delhi: Sage.

'Akau'ola, (1999), "Tongan Whaling Traditions," in *Issues Related to Indigenous Whaling: Tonga*, Freeman, M. (ed.), Brentwood Bay, Canada: World Council of Whalers.

Amoore, L., and Langley, P. (2004), "Ambiguities of Global Civil Society," *Review of International Studies*, vol. 30, pp. 89–110.

Anderson, B. (1983), *Imagined Communities*, London: Verso.

Anderson, J. M. (1981), *Ecology for Environmental Sciences*, London: Edward Arnold.

Anderson, R. (2001), Makah Hunt for Whales May Enter New Waters, *Seattle Times*, 14/01/01.

Andresen, S. (2001), "The Whaling Regime: 'Good' Institution but 'Bad' Politics?," in *Towards a Sustainable Whaling Regime*, Friedheim, R. L. (ed.), Seattle, WA, and London: University of Washington Press and Edmonton: Canadian Circumpolar Institute.

Antunes, J. (2000), "Regime Effectiveness, Joint Implementation and Climate Change Policy," in *Climate Change and American Foreign Policy*, Harris, P. G. (ed.), New York: St. Martin's Press.

Arlov, T. B. (1993), "Whaling and Sovereignty: The Role of Whaling in the Struggle for Supremacy over Svalbard (Spitsbergen)," in *Whaling and History: Perspectives on the Evolution of the Industry*, Basberg, B. L., Ringstad, J. E., and Wexelsen, E. (eds.), Sandefjord, Norway: Sandefjordmuseene.

Aron, W. (2001), "Science and the IWC," in *Towards a Sustainable Whaling Regime*, Friedheim, R. L. (ed.), Seattle, WA, and London: University of Washington Press and Edmonton: Canadian Circumpolar Institute.

Aron, W., Burke, W., and Freeman, M. R. (2001), "The Whaling Issue," in *Marine Policy*, vol. 24, pp. 179–191.

Aronowitz, S. (1988), *Science as Power: Discourse and Ideology in Modern Society*, Minneapolis: University of Minnesota Press.

Asahi Shinbun (evening), 29/06/1992.

Ásgrímsson, H. (1997), Opening Address, International Conference on Whaling in the North Atlantic, Reykjavik, March 1, 1997, http://www.highnorth.no/Library/Hunts/Iceland/ic-po-on.htm, accessed 07/11/07.

Associated Press briefing, 10/04/2002, *Daily Telegraph*.

Austin, J. (1962), *How To Do Things with Words*, Oxford: Clarendon.

Bachelard, G. (1947), *L'eau et les rêves: Essais sur l'imagination de la matière*, Paris: J. Corti.

Baldwin, D. (2006), "Power and International Relations," in *Handbook of International Relations*, Carslnaes, W., Risse, T., and Simmons, B. A. (eds.), London, Thousand Oaks, and New Delhi: Sage.

Balibar, E. (1991), "Is There a Neo-Racism?," in Balibar, E., and Wallerstein, I. (eds.), *Race, Nation, Class*, London: Verso.

Barber, M. (1999), Pacific Gray Whales Have Made a Strong Comeback, *Seattle Post Intelligencer*, 18/03/1999.

Barkawi, T., and Laffey, M. (2002), "Retrieving the Imperial: Empire and International Relations," *Millennium*, vol. 31, no. 1, pp. 109–128.

Barker, C. J. (2000), *International Law and International Relations*, London and New York: Continuum.

Barnett, M., and Duvall, R. (2005), "Power in International Politics," *International Organization*, vol. 59, pp. 39–75.

Barry, J., and Eckersley, R. (eds.) (2005), *The State and the Global Ecological Crisis*, Cambridge, MA: MIT Press.

Barsh, R. L. (2001), "Food Security, Food Hegemony and Charismatic Animals," in *Towards a Sustainable Whaling Regime*, Friedheim, R. L. (ed.), Seattle, WA, and London: University of Washington Press and Edmonton: Canadian Circumpolar Institute.

Bartelson, J. (1995), *A Genealogy of Sovereignty*, Cambridge: Cambridge University Press.

Batisse, M. (1982), "The Biosphere Reserve: A Tool for Environmental Conservation and Management," *Environmental Conservation*, vol. 9, no. 2, pp. 101–111.

Baudrillard, J. (1970), *The Consumer Society: Myths and Structures*, (1998 English ed.), London: Sage.

Baudrillard, J. (1972), *For A Critique of the Political Economy of the Sign* (1981, English ed.), St. Louis: Telos Press.

Baudrillard, J. (1975), *Le miroir de la production ou l'illusion critique du matérialisme historique*, Paris: Galilée.

Baudrillard, J. (1988a), "Simulacra and Simulations," in *Jean Baudrillard: Selected Writings*, Poster, M. (ed.), Oxford: Blackwell/Polity Press.

Baudrillard, J. (1988b), "System of Objects," in *Jean Baudrillard: Selected Writings*, Poster, M. (ed.), Oxford: Blackwell/Polity Press.

Baudrillard, J. (1993), *Symbolic Exchange and Death*, London: Sage.

Bayet, F. (1994), "Overturning the Doctrine: Indigenous People and Wilderness—Being Aboriginal in the Environmental Movement," *Social Alternatives*, vol. 19, no. 4, pp. 1–19.

Beck, U. (1992), *Risk Society*, London, Thousand Oaks, and New Delhi: Sage.

Beck, U. (2000), *The Brave New World of Work*, Cambridge: Polity Press.

Beck, U., Giddens, A., and Lash, S. (eds.) (1994), *Reflexive Modernization: Politics, Tradition and Aesthetics in the Modern Social Order*, Cambridge: Polity Press.

Belich, J. (1996), *Making Peoples: A History of New Zealanders*, Honolulu: University of Hawaii Press.

Bell, D. (1979), *The Cultural Contradictions of Capitalism*, London: Heinemann.

Benton, L. M., and Short, J. R. (1999), *Environmental Discourse and Practice*, Oxford: Blackwell.

Benton, L. M., and Short, J. R. (2000), *Environmental Discourse and Practice: A reader*, Oxford: Blackwell.

Benton, T. (1989), "Marxism and Natural Limits: An Ecological Critique and Reconstruction," *New Left Review*, vol. 178, pp. 51–86.

Berenskoetter, F. (ed.) (2008), *Power in World Politics*, London and New York: Routledge.

Betsil, M. (2000), "The United States and the Evolution of International Climate Change Norms," in *Climate Change and American Foreign Policy*, Harris, P. G. (ed.), New York: St. Martin's Press.

Beverton, R. J. H., and Holt, S. J. (1957), *On the Dynamics of Exploited Fish Populations*, London: H.S.M.O.

Bially Mattern, J. (2005), *Ordering International Politics: Identity, Crisis and Representational Force*, London and New York: Routledge.

Biernecki, R. (1995), *The Fabrication of Labor*, Berkeley, CA: University of California Press.

Birnie, P. (1985), "The Role of Developing Countries in Nudging the IWC form Regulating Whaling to Encouraging Non-Consumptive Uses of Whales," *Ecology Law Quarterly*, vol. 12, no. 4, pp. 937–974.

Black, R. (2007), "Whaling Fights: Everyone's A Winner," BBC News online, 01/06/07, http://news.bbc.co.uk/2/hi/science/nature/6711885.stm, accessed 10/03/08.

Blichfeldt, G. (1994), "Bigger than Whales," in *11 Essays on Whales and Man*, High North Alliance (ed.), Reine, Norway: HNA.

Boli, J., and Thomas, G. M. (eds.) (1999), *Constructing World Culture: International Non-Governmental Organizations since 1875*, Stanford: Stanford University Press.

Bonner, W. N. (1980), *Whales*, Poole, Dorset: Blandford Press.

Booth, K., and Smith, S. (eds.) (1995), *International Relations Theory Today*, Cambridge: Polity Press.

Bourdieu, P. (1980), *Le Sens Pratique*, Paris: Editions de Minuit.

Bourdieu, P. (1983), *Ce que parler veut dire: L'économie des échanges linguistiques*, Paris: Fayard.

Bourdieu, P. (1991), *Language and Symbolic Power*, Thompson, J. (ed.), Cambridge: Cambridge University Press/Polity Press.

Bourdieu, P. (1998), *La domination masculine*, Paris: Seuil/Points.

Bourdieu, P. (2001), *Langage et pouvoir symbolique*, Paris: Editions Fayard.

Boudieu, P. (2002a), "Quelques proprietés des champs," in *Questions de Sociology*, Paris: Editions de Minuit.

Boudieu, P. (2002b), "Comment être sportif?," in *Questions de Sociology*, Paris: Editions de Minuit.

Bourdieu, P. (2002c), "Dévoiler les ressorts du pouvoir," in *Interventions*, Paris: Antigone.

Bourdieu, P. (2003), *Méditations pascaliennes*, Paris: Editions du Seuil.

Boyle, A. E., and Birnie, P. W. (1992), *International Law and the Environment*, Oxford: Clarendon.

Brading, D. A. (1984), "Bourbon Spain and Its American Empire," in *The Cambridge History of Latin America, Vol. 1*, Bethell, L. (ed.), Cambridge: Cambridge University Press.

Brenton, T. (1994), *The Greening of Machiavelli: The Evolution of International Environmental Politics*, London and Earthscan: Royal Institute for International Affairs.

Brookings Institution (1998), "National Environmental Policy at 30 Years," http://www.brook.edu/dybdocroot/pressroom/transcripts/19981013a.htm., accessed 02/05/03.

Brown, C. (1992), *International Relations Theory: New Normative Approaches*, London and New York: Harvester Wheatsheaf.

Brown, G., and Yule, G. (1983), *Discourse Analysis*, Cambridge: Cambridge University Press.

Brown, P. (2002), "Japanese Branch Outrages WWF with Their Whaling Plea," *The Guardian*, 02/04/02.

Bruntland, G. H. (1993), "Whales and Sustainable Ocean Management," *The Network*, April 1993, http://www.highnorth.no/Library/Policies/National/wh-an-su.htm, accessed 07/11/07.

Brydon, A. (1991), *The Eye of the Guest: Icelandic National Discourse and the Whaling Issue*, unpublished PhD dissertation, McGill University.

Brydon, A. (1996), "Whale-Siting: Spatiality in Icelandic Nationalism," in *Images of Contemporary Iceland: Everyday Lives and Global Contexts*, Palsson, G., and Durrenberger, E. P. (eds.), Iowa City: University of Iowa Press.

Bull, H. (1977), *The Anarchical Society: A Study of Order in World Politics*, London: Palgrave.

Burch, K. (1994), "The 'Properties' of the State System and Global Capitalism," in *The Global Economy as Political Space*, Rosow, S., Inayatullah, N., and Rupert, M. (eds.), Boulder, CO, and London: Lynne Rienner.

Burch, K. (1998), *"Property" and the Making of the International System*, Boulder, CO, and London: Lynne Rienner.

Burke, W. (2001), "A New Whaling Agreement and International Law," in *Towards a Sustainable Whaling Regime*, Friedheim, R. L. (ed.), Seattle, WA, and London: University of Washington Press and Edmonton: Canadian Circumpolar Institute.

Butler, J. (1997), *The Psychic Life of Power: Theories in Subjection*, Stanford: Stanford University Press.

Butler, J. (2004), *Precarious Life: The Powers of Mourning and Violence*, London and New York: Verso.

Butler, J., Laclau, E., and Zizek, S. (2000), *Contingency, Hegemony, Universality: Contemporary Dialogues on the Left*, London and New York: Verso.

Calsamiglia, H., and Lopez-Ferrero, C. (2003), "Role and Position of Scientific Voices: Reported Speech in the Media," *Discourse Studies*, vol. 5, no. 2, pp. 147–173.

Campbell, D. (1992), *Writing Security: United States Foreign Policy and the Politics of Identity*, Manchester: Manchester University Press.

Campbell, D. (1994), "Foreign Policy and Identity: Japanese 'Other'/American 'Self,'" in *The Global Economy as Political Space*, Rosow, S., Inayatullah, N., and Rupert, M. (eds.), Boulder, CO, and London: Lynne Rienner.

Caron, D. D. (1995), "The International Whaling Commission and the North Atlantic Marine Mammals Commission: The Institutional Risks of Coercion in Consensual Structures," *American Journal of International Law*, vol. 89, no. 1, pp. 154–174.

Carter, N. (2001), *The Politics of the Environment: Ideas, Activism, Policies*, Cambridge: Cambridge University Press.

Carver, T. (2002), "Discourse Analysis and the 'Linguistic Turn,'" *European Political Science*, vol. 2, no. 1, pp. 50–53.

Carver, T., and Hyvärinen, M. (eds.) (1997), *Interpreting the Political: New Methodologies*, London: Routledge.

Castells, M. (1996), *The Rise of the Network Society, Vol. 1*, Oxford: Blackwell.

Cawthorn, M. C. (1999), "The Changing Face of New Zealand's Whaling Policy," in *Whaling and Anti-whaling Movement*, Tokyo: ICR.

Central Intelligence Agency (1956), "National Intelligence Survey: Antarctica. Chapter V—Political" (declassified October 26, 1989).

Chafetz, G., Spirtas, M., and Frankel, B. (1999), "Tracing the Influence of Identity," in *The Origins of National Interest*, Chafetz, G., Spirtas, M., and Frankel, B. (eds.), London and Portland, OR: Frank Cass.

Chandler, D. (2007), *Semiotics: The Basics*, London and New York: Routledge.

Chapman, C., Kumar, K., Fraser, C., and Gaber, I. (1997), *Environmentalism and the Mass Media: The North–South Divide*, London and New York: Routledge.

Charnovitz, S. (1995), "Environmental Trade Sanctions and the GATT: An Analysis of the Pelly Amendment on Foreign Environmental Practices," in *Additional Essays on Whales and Man*, High North Alliance (eds.), Reine, Norway: HNA.

Checkel, J. T. (1998), "The Constructivist Turn in International Relations Theory," *World Politics*, vol. 50, no. 2, pp. 324–348.

Checkel, J. T. (2005), "International Institutions and Socialization in Europe: Introduction and Framework," *International Organization*, vol. 59, no. 4, pp. 801–826.

Cherfas, J. (1988), *The Hunting of the Whale: A Tragedy that Must End*, London: The Bodley Head.

Christian, P. (2000), "Nelson Whale Hui Slammed," *The Christchurch Press*, 20/06/2000.

Christoffersen, N. (1994), "Whaling, Cultural Diversity and Sustainable Communities," in *11 Essays on Whales and Man*, High North Alliance (eds.), Reine, Norway: HNA.

Clegg, S. (2000), "Power and Authority, Resistance and Legitimacy," in *Power in Contemporary Politics*, Goverde, H., Cerny, P. H., Haugaard, M., and Lentner, H. (eds.). London, Thousand Oaks, and New Dehli: Sage.

Clegg, S. R. (1989), *Frameworks of Power*, London, Thousand Oaks, and New Delhi: Sage.

Conca, K. (2002), "Consumption and Environment in a Global Economy," in *Confronting Consumption*, Princen, T., Maniantes, M., and Conca, K. (eds.), Cambridge, MA: MIT Press.

Connelly, S. (2007), "Mapping Sustainable Development as a Contested Concept," *Local Environment*, vol. 12, no. 3, pp. 259–278.

Connolly, W. (1993), *The Terms of Political Discourse*, Pinceton, NJ: Princeton University Press.

Connolly, W. (1989), "Identity and Difference in Global Politics," in *International/Intertextual Relations: Postmodern Readings of World Politics*, Der Derian, J., and Shapiro, M. (eds.), Lexington: Lexington Books.

Coutts, P. J. F. (1976), "An Approach to the Investigation of Colonial Settlement Patterns: Whaling in Southern New Zealand," *World Archaeology*, vol. 7, no. 3, pp. 291–305.

Cox, R. (1993), "Gramsci, Hegemony and International Relations: An Essay in Method," in *Gramsci, Historical Materialism and International Relations*, Gill, S. (ed.), Cambridge: Cambridge University Press.

Creighton, M. S. (1995), *Rites and Passages: The Experience of US Whaling*, Cambridge: Cambridge University Press.

D'Amato, A., and Chopra, S. K. (1991), "Whales: Their Emerging Right to Life," *American Journal of International Law*, vol. 85, no. 1, pp. 21–62.

Danet, B. (1985), "Legal Discourse," in *Handbook of Discourse Analysis, Vol. 1*, Dijk, T. V. (ed.), London: Academic Press.

Darier, E. (ed.) (1999), *Discourses of the Environment*, Oxford: Blackwell.

Davis, L. E., and Gallman, R. E. (1993), "American Whaling, 1820–1900: Dominance and Decline," in *Whaling and History: Perspectives on the Evolution of the Industry*, Basberg, B. L., Ringstad, J. E., and Wexelsen, E. (eds.), Sandefjord, Norway: Sandefjordmuseene.

Day, D. (1992), *The Whale War*, London: Grafton.

Deleuze, G. (1986), *Foucault*, Paris: Editions de minuit.

DeLuca, K. M. (1999), *Image Politics: The New Rhetoric of Environmental Activism*, New York: Guilford Press.

Der Derian, J., and Shapiro, M. (eds.) (1989), *International/Intertextual Relations: Postmodern Readings of World Politics*, Lexington: Lexington Books.

Derrida, J. (1972), *Positions* (1981, English ed.), Chicago, IL: University of Chicago Press.

DeSombre, E. R. (2000), *Domestic Sources of International Environmental Policy: Industry, Environmentalists and U.S. Power*, Cambridge, MA, and London: MIT Press.

DeSombre, E. R. (2001), "Distorting Global Governance: Membership, Voting and the IWC," in *Towards a Sustainable Whaling Regime*, Friedheim, R. L. (ed.), Seattle, WA, and London: University of Washington Press and Edmonton: Canadian Circumpolar Institute.

DeSombre, E. R. (2002), *The Global Environment and World Politics*, London and New York: Continuum.

Dessler, D. (1989), "What Is at Stake in the Agent-Structure Debate," *International Organization*, vol. 43, no. 3, pp. 441–473.

Diez, T. (2001), "Europe as a Discursive Battleground: Discourse Analysis and European Integration Studies," *Cooperation and Conflict*, vol. 36, no. 1, pp. 5–38.

Dijk, T. V. (ed.) (1985a), *Handbook of Discourse Analysis, Vols. I–IV*, London: Academic Press.

Dijk, T. V. (ed.) (1985b), *Discourse and Communication: New Approaches to the Analysis of Mass Media Discourse and Communication*, New York: De Gruyter.

Donovan, G. P. (1989), "Report of the Special Meeting of the Scientific Committee on Planning for a Comprehensive Assessment of Whale Stocks," in *The Comprehensive Assessment of Whale Stocks*, Cambridge, UK: IWC.

Donovan, G. P. (1992), "The International Whaling Commission: Given its Past, Does It Have a Future?," in *Proceedings of a Symposium on Whales: Biology–Threat–Conservation*, Symoens, J. J. (ed.), pp. 23–44, Brussels: Royal Academy of Overseas Science.

Donovan, G. P. (1995), "The International Whaling Commission and the Revised Management Procedure," in *Additional Essays on Whales and Man*, High North Alliance (eds.), Reine, Norway: HNA.

Donovan, G. P. (1999), Editorial to *Journal of Cetacean Research and Management*, vol. 1, pp. 28–44.

Doran, P. (1995), "Earth, Power, Knowledge: Towards a Critical Global Environmental Politics," in *Boundaries in Question*, Macmillan, J., and Linklater, A. (eds.), London and New York: Pinter.

Dorney, S. (2005), Nauru Defends Whaling Vote, ABC News online, 28/06/05, http://www.abc.net.au/news/newsitems/200506/s1402449.htm, accessed 17/10/07.

Douglas, M. (2002), *Food in the Social Order*, London and New York: Routledge.

Dryzek, J. (1997), *The Politics of the Earth: Environmental Discourses*, Oxford: Oxford University Press.

Dryzek, J. S., Clark, M. L., and McKenzie, G. (1989), "Subject and System in International Interaction," *International Organization*, vol. 43, no. 3, pp. 475–503.

Dryzek, J. S., and Schlossberg, D. (eds.) (2005), *Debating the Earth: The Environmental Politics Reader*, Oxford: Oxford University Press.

Durning, A. (1992), *How Much Is Enough? The Consumer Society and the Future of the Earth*, New York: W.W. Norton.

Eco (2001a). Yen Diplomacy: Bribes, Lies and Audiotapes, vol. LIII, no. 1, 23/07/01.

Eco (2001b). New Zealand Slams "Blatant" Vote-Buying, vol. LIII, no. 4, 26/07/01.

Edkins, J. (1999), *Postructuralism and International Relations: Bringing the Political Back In*, Boulder, CO, and London: Lynne Rienner.

Einarsson, N. (1993), "All Animals Are Equal, but Some Are Cetaceans: Conservation and Culture Conflict," in *Environmentalism: The View from Anthropology*, Milton, K. (ed.), London and New York: Routledge.

Elliot, L. M. (ed.) (1994), *International Environmental Politics: Protecting the Antarctic*, London and New York: St. Martin's Press.

Ellis, R. (1992), *Men and Whales*, London: Robert Hale Ltd.

Engle-Merry, S. (2000), *Colonizing Hawai'i*, Princeton, NJ: Princeton University Press.

Epstein, C. (2003), "World Wide Whale: Globalisation and a Dialogue of Cultures?," *Cambridge Review of International Affairs*, vol. 16, no. 2, pp. 309–322.

Epstein, C. (2004), "A Fractured Hegemony? The US Position in the International Politics of Whaling," paper presented at the International Studies Association Annual Conference, Montreal, Canada.

Epstein, C. (2005), "Knowledge and Power in Global Environmental Activism," *International Journal of Peace Studies*, vol. 10, no. 1, pp. 47–67.

Epstein, C. (2006), "The Making of Global Environmental Norms: Endangered Species Protection," *Global Environmental Politics*, vol. 6, no. 2, pp. 32–55.

Epstein, C. (2007), "Guilty Bodies, Productive Bodies, Destructive Bodies: Crossing the Biometric Borders," *International Political Sociology*, vol. 1, no. 2, pp. 149–164.

Escobar, A. (1994), "Welcome to Cyberia: Notes on the Anthropology of Cyberculture," *Current Anthropology*, vol. 35, no. 3, pp. 211–231.

Escobar, A. (1996), "Construction Nature: Elements for a Post-structuralist Political Ecology," *Futures*, vol. 28, no. 4, pp. 325–343.

Escobar, A. (2001), "Culture Sits in Places: Reflections on Globalism and Subaltern Strategies of Localization," *Political Geography*, vol. 20, pp. 139–174.

Fairclough, N. (1992), *Discourse and Social Change*, Cambridge: Polity Press.

Fairclough, N. (1995), *Media Discourse*, London, New York, Sydney, and Auckland: Arnold.

Falk, R. (1989), "Introduction: Preserving Whales in a World of Sovereign States," *Denver Journal of International Law and Policy*, vol. 17, pp. 249–254.

Featherstone, M. (1991), *Consumer Culture and Postmodernism*, London, Newbury Park, and New Delhi: Sage.

Ferguson, J. (1994), *The Anti-politics Machine: "Development," Depoliticization and Bureaucratic Power in Lesotho*, Minneapolis: University of Minnesota Press.

Ferry, L. (1992), *Le nouvel ordre écologique: L'arbre, l'animal et l'homme*, Paris: Grasset et Fasquelle.

Fiji South Pacific Consumer Protection Programme (1999), "What Is the Problem With Mutton Flaps?," in *Issues Related to Indigenous Whaling: Tonga*, Freeman, M. (ed.), Brentwood Bay, Canada: World Council of Whalers.

Finnemore, M., and Sikkink, K. (1998), "International Norms Dynamics and Political Change," *International Organization*, vol. 52, no. 4, pp. 887–917.

Fisher, F., and Hajer, M. H. (1999), *Living with Nature: Environmental Politics as Cultural Discourse*, Oxford: Oxford University Press.

Flockhart, T. (2006), "Complex Socialization: A Framework for the Study of State Socialisation," *European Journal of International Relations*, vol. 12, no. 1, pp. 89–118.

Food and Agriculture Organization (2001), *The State of Food and Agriculture: Economic Impacts of Transboundary Plant Pests and Animal Diseases*, Rome: FAO Publications.

Food and Agriculture Organization (1978–81), *Mammals in the Sea, Vols. I–IV*, Rome: FAO Publications/FAO Fisheries Series.

Forgacs, D. (1988), *A Gramsci Reader*, London: Lawrence and Wishart.

Foucault, M. (1967), *Madness and Civilization: A History of Insanity in the Age of Reason*, London: Tavistock.

Foucault, M. (1970), *The Order of Things*, London: Tavistock.

Foucault, M. (1972), *The Archaeology of Knowledge*, London: Tavistock.

Foucault, M. (1975), *Surveiller et punir: Naissance de la prison*, Paris: Gallimard.

Foucault, M. (1977), *Language, Counter-Memory, Practice: Selected Essays and Interviews*, Bouchard, D. F., and Simons, S. (eds.), Oxford: Blackwell.

Foucault, M. (1980), "Truth and Power," in *Power/Knowledge: Selected Interviews And Other Writings (1972–1977)*, Gordon, C. (ed.), London: Tavistock.

Foucault, M. (1981), "The Order of Discourse," in *Untying the Text: A Post-structuralist Reader*, Young, R. (ed.), London: Routledge and Boston: Paul Kegan.

Foucault, M. (1982), *Michel Foucault: Beyond Structuralism and Hermeneutics*, Dreyfus, H. L., and Rabinow, P. (eds.), Brighton: Harvester.

Foucault, M. (1988a), *Politics, Philosophy, Culture: Interviews and Other Writings 1977–1984*, Kritzman, L. D. (ed.), London and New York: Routledge.

Foucault, M. (1988b), "Truth, Power, Self: An Interview with Michel Foucault—October 25th 1982," in *Technologies of the Self: A Seminar with Michel Foucault*, Martin, L., et al. (eds.), London: Tavistock.

Foucault, M. (1990), *History of Sexuality, Vol. 1*, translated by Robert Hurley, New York: Random House/Vintage.

Foucault, M. (1991), "Questions of Method," in *The Foucault Effect: Studies in Governmentality*, Burchell, G., Gordon, C., and Miller, P. (eds.), Chicago: University of Chicago Press.

Foucault, M. (2001), "La Governmentalité," in *Dits et Ecrits, Vol. II (1976–1988)*, Defert, D., Ewald, F., and Lagrange, J. (eds.), Paris: Gallimard.

Foucault, M. (2003), *Society Must Be Defended: Lectures at the Collège de France 1975–1976*, New York: Picador.

Fowler, R. (1985), "Power," in *Handbook of Discourse Analysis, Vol. IV*, Dijk, T. V. (ed.), London: Academic Press.

Fowler, R., Hodge, B., Kress, G., and Trew, T. (eds.) (1979), *Language and Control*, London, Boston, and Henley: Routledge and Paul Kegan.

Francks, P., Boestel, J., and Kim, C. H. (1999), "The State and Agricultural Adjustment in Japan," in *Agriculture and Economic Development in East Asia*, Francks, P., Boestel, J., and Kim, C. H. (eds.), London and New York: Routledge.

Frank, D. J., Hironaka, A., Meyer, J. W., Schofer, E., and Tuma, N. B. (1999), "The Rationalisation and Organization of Nature in World Culture," in *Constructing World Culture: International Non-Governmental Organizations since 1875*, Boli, J., and Thomas, G. (eds.), Stanford: Stanford University Press.

Freeman, M. R. (1993), "The International Whaling Commission, Small-Type Whaling, and Coming to Terms with Subsistence," *Human Organization*, vol. 52, no. 3, pp. 243–251.

Freeman, M. R. (1994), "Gallup on Public Attitudes to Whales and Whaling," in *11 Essays on Whales and Man*, High North Alliance (eds.), Reine, Norway: HNA.

Freeman, M. R. (1995), "Economy, Equity and Ethics: Current Perspectives in Wildlife Management in the North," *Human Ecology: Issues in the North*, vol. 3, pp. 3–12.

Freeman, M. R. (1996), "Polar Bears and Whales: Contrasts in International Wildlife Regimes," *Issues in the North*, vol. 1, no. 40, pp. 174–181.

Freeman, M. R. (ed.) (2000), *Endangered Peoples of the Arctic: Struggles to Survive and Thrive*, Westport, CT, and London: Greenwood Press.

Freeman, M. R. (2001), "Is Money the Root of the Problem? Cultural Conflict in the IWC," in *Towards a Sustainable Whaling Regime*, Friedheim, R. L. (ed.), Seattle, WA, and London: University of Washington Press and Edmonton: Canadian Circumpolar Institute.

Freeman, M. R., and Kellert, S. R. (1992), *Public Attitude to Whales: Result of a Six Country Survey*, Alberta, Canada: Canadian Circumpolar Institute and Yale, CT: School of Forestry and Environmental Studies.

Friedheim, R. L. (2001a), "Negotiating in the IWC Environment," in *Towards a Sustainable Whaling Regime*, Friedheim, R. L. (ed.), Seattle, WA, and London: University of Washington Press and Edmonton: Canadian Circumpolar Institute.

Friedheim, R. L. (2001b), "Fixing the Whaling Regime: A Proposal," in *Towards a Sustainable Whaling Regime*, Friedheim, R. L. (ed.), Seattle, WA, and London: University of Washington Press and Edmonton: Canadian Circumpolar Institute.

Friedheim, R. L. (ed.) (2001), *Towards a Sustainable Whaling Regime*, Seattle, WA, and London: University of Washington Press and Edmonton: Canadian Circumpolar Institute.

Friends of the Earth (1978), *The Whale Manual*, FOE Publications.

Gambell, R. (1983), "Bowhead Whales and Alaskan Eskimos: A Problem of Survival," *Polar Record*, vol. 21, no. 134, pp. 467–473.

Gambell, R. (1989), "Interview with Ray Gambell," *The Pilot*, vol. 3.

Gambell, R. (1995), "Management of Whaling in Coastal Communities," in *Whales, Seals, Fish and Man*, Blix, A. S., Walløe, L., and Ulltang, Ø. (eds.), Amsterdam: Elsevier.

Gambell, R. (1997a), "Recent Developments in the IWC Aboriginal Subsistence Whaling Category," in *Whaling in the North Atlantic: Economic and Political Perspectives*, Petursdottir, G. (ed.), Reykjavik: Fisheries Research Institute.

Gambell, R. (1997b), "The International Whaling Commission Today," in *Whaling in the North Atlantic: Economic and Political Perspectives*, Petursdottir, G. (ed.), Reykjavik: Fisheries Research Institute.

Garnham, N. (1990), *Capitalism and Communication: Global Culture and the Economics of Information*, London, Newbury Park, and New Delhi: Sage.

Gellner, E. (1970), "Concepts and Society," in *Rationality*, Wilson, B. R. (ed.) (1984, 2nd ed.), Oxford: Blackwell.

Giddens, A. (1979), *Central Problems in Social Theory*, London: Macmillan.

Giddens, A. (1990), *The Consequences of Modernity*, Cambridge: Polity Press/ Blackwell.

Giddens, A. (1991), *Modernity and Self-Identity*, Stanford: Stanford University Press.

Gill, S. (1993), "Gramsci and Global Politics: Towards a Post-hegemonic Research Agenda," in *Gramsci, Historical Materialism and International Relations*, Gill, S. (ed.), Cambridge: Cambridge University Press.

Gill, S. (2003), *Power and Resistance in the New World Order*, New York and London: Palgrave.

Gillespie, A. (2001), "The Evolving Bi-cultural Relationship with the Management of Cetaceans in New Zealand," *Asia Pacific Journal of Environmental Law*, vol. 6, no. 1, pp. 13–46.

Goldstein, J. (1989), "The Impact of Ideas on Trade Policy: The Origins of U.S. Agricultural and Manufacturing Policies," *International Organization*, vol. 43, no. 1, pp. 31–71.

Goldstein, J., and Keohane, R. O. (eds.) (1993), *Ideas and Foreign Policy: Beliefs, Institutions and Political Change*, Ithaca and London: Cornell University Press.

Goverde, H., Cerny, P. H., Haugaard, M., and Lentner, H. (eds.) (2000), *Power in Contemporary Politics: Theories, Practises, Globalizations*, London, Thousand Oaks, and New Dehli: Sage.

Gramsci, A. (1971), *Selections from Prison Notebooks*, Hoare, Q., and Nowell Smith, G. (eds.), London: Lawrence and Wishart.

Green, O. (1996), "Environmental Regimes: Effectiveness and Implementation Review," in *The Environment and International Relations*, Vogler, J., and Imber, M. (eds.), London and New York: Routledge.

Greenpeace (2004a), *Annual Report*, http://www.greenpeace.org/international_en/, accessed 12/11/2004.

Greenpeace (2004b), "Donations," https://www.greenpeace.com/donation/, accessed 12/11/2004.

Greimel, H. (2002), Most Japanese Support Commercial Whaling According to Survey, Associated Press, 17/03/02.

Grove-White, R. (1993), "Environmentalism: A New Moral Discourse for a Technological Society?," in *Environmentalism: The View from Anthropology*, Milton, K. (ed.), London and New York: Routledge.

Guha, R. (2000), *Environmentalism: A Global History*, New York: Longman.

Gulland, J. (1988), "The Management Regime for Living Resources," in *The Antarctic Legal Regime*, Joyner, C. C., and Chopra, S. K. (eds.), Dordrecht, Boston, and London: Martinus Nijhoff.

Guzzini, S. (2000), "A Reconstruction of Constructivism in International Relations," *European Journal of International Relations*, vol. 6, no. 2, pp. 147–182.

Guzzini, S. (2008), "The Concept of Power: A Constructivist Analysis," in *Power in World Politics*, Berenskoetter, F. (ed.), London and New York: Routledge.

Haas, E. B. (1980), "Why Collaborate? Issue-Linkage and International Regimes," *World Politics*, vol. 32, no. 3, pp. 357–405.

Haas, P. (1992), "Introduction: Epistemic Communities and International Policy Coordination," *International Organization* (Special Issue on Knowledge, Power and International Policy Coordination), vol. 46, no. 1, pp. 1–35.

Haas, P. (2004), "When Does Power Listen To Truth? A Constructivist Approach to the Policy Process," *Journal of European Public Policy*, vol. 11, no. 4, pp. 569–592.

Haas, P., Keohane, R. O., and Levy, M. (eds.) (1993), *Institutions for the Earth: Sources of Effective International Environmental Protection*, Cambridge, MA, and London: MIT Press.

Habermas, J. (1971), *Knowledge and Human Interest*, Boston: Beacon Press.

Habermas, J. (1987), *The Philosophical Discourse of Modernity*, Cambridge: Polity Press.

Haggard, S., and Simmons, B. A. (1987), "Theories of International Regimes," *International Organization*, vol. 41, no. 3, pp. 491–517.

Hajer, M. (2002), "Discourse Analysis and the Study of Policy Making," *European Political Science*, vol. 2, no. 1, pp. 61–65.

Hajer, M. A. (1995), *The Politics of Environmental Discourse: Modernization and the Policy Process*, Oxford: Oxford University Press/Clarendon.

Hall, S., and Gieben, B. (1991), *Formations of Modernity*, Cambridge: Polity Press/Open University.

Hannesson, R. (2006), *The Privatization of the Oceans*, Cambridge, MA: MIT Press.

Hannigan, J. A. (1995), *Environmental Sociology: A Social Constructionist Perspective*, London and New York: Routledge.

Hansen, L. (2006), *Security As Practice: Discourse Analysis and the Bosnian War*, London and New York: Routledge.

Haraway, D. J. (1991), *Simians, Cyborgs and Women: The Reinvention of Nature*, London and New York: Routledge.

Hardt, M., and Negri, A. (2001), *Empire*, Cambridge, MA, and London: Harvard University Press.

Harré, R., Brockmeier, J., and Mühlhäusler, P. (eds.) (1999), *Greenspeak: A Study of Environmental Discourse*, London, Thousand Oaks, and New Delhi: Sage.

Harries-Jones, P. (1993), "Between Science and Shamanism: The Advocacy of Environmentalism in Toronto," in *Environmentalism: The View from Anthropology*, Milton, K. (ed.), London and New York: Routledge.

Harris, P. G. (ed.) (2000), *Climate Change and American Foreign Policy*, New York: St. Martin's Press.

Hawkes, D. (2003), *Ideology*, London and New York: Routledge.

Hay, C., and Rosamond, B. (2002), "Globalization, European Integration and Discursive Construction of Economic Perspectives," *Journal of European Public Policy*, vol. 9, no. 2, pp. 147–167.

Herman, D. J. (2001), *Hunting and the American Imagination*, Washington and London: Smithsonian Institute.

Hess, B. (1999), *Gift of the Whale: The Inupiat Bowhead Hunt, a Sacred Tradition*, Seattle, WA: Sasquatch Books.

High North Alliance (ed.) (1994a), *11 Essays on Whales and Man*, Reine, Norway: HNA.

High North Alliance (1994b), "US Chief Scientist Rebuked and Rewarded," *High North Alliance News*, no. 9, (12/12/94).

High North Alliance (ed.) (1995), *Additional Essays in Whales and Man*, Reine, Norway: HNA.

Hiraguchi, T. (2003), "Prehistoric and Protohistoric Whaling and Diversity in Japanese Foods," in *Traditional Whaling Summit in Nagato*, Institute for Cetacean Research (ed.), Tokyo and Nagato: Institute for Cetacean Research and Nagato City.

Hoel, A. H. (1998), "Political Uncertainty in International Fisheries Management," *Fisheries Research*, vol. 37, pp. 239–250.

Hollis, M., and Smith, S. (1990), *Explaining and Understanding International Relations*, Oxford: Oxford University Press/Clarendon.

Holt, S. (2000), "International Aspects of Whale-Watching," *Various Aspects of Whale-Watching*, Report from an ICRAM/IFAW workshop.

Holzscheiter, A. (2005), "Discourse as Capability: Non-State Actors' Capital in Global Governance," *Millenium*, vol. 33, no. 3, pp. 723–746.

House of Commons (1991), Record of Debates in the House of Commons, "Whales," columns 35–42, 03/06/1991.

Howarth, D. (1995), "Discourse Theory," in *Theory and Methods in Political Science*, Marsh, D., and Stoker, G. (eds.), London: Macmillan.

Howarth, D. (2000), *Discourse*, Buckingham and Philadelphia: Open University Press.

Hoyt, E. (2000), *Report of the Closing Workshop to Review Various Aspects of Whale-Watching*, IFAW/ICRAM.

Hoyt, E. (2001), *Whale-watching Report 2001*, IFAW/UNEP.

Hunter, R. (1980), *The Greenpeace Chronicles*, London: Picador.

Hurrell, A., and Kingsbury, B. (1992), *The International Politics of the Environment: Actors, Interests, Institutions*, Oxford: Clarendon Press.

Hutton, J., and Dickson, B. (eds.) (2000), *Endangered Species, Threatened Convention: The Past, Present and Future of CITES*, London: Earthscan.

Huysmans, J. (1998), "Security! What Do You Mean? From Concept to Thick Signifier," *European Journal of International Relations*, vol. 4, no. 2, pp. 226–255.

Ingold, T. (1993), "Globes and Spheres: The Topology of Environmentalism," in *Environmentalism: The View from Anthropology*, Milton, K. (ed.), London and New York: Routledge.

Institute for Cetacean Research (1988a), "History of Whaling in Japan," in *Small Type Coastal Whaling in Japan: Report of an International Workshop*, Edmonton, Canada: Boreal Institute for Northern Studies.

Institute for Cetacean Research (1988b), "Contemporary Small Type Coastal Whaling," in *Small Type Coastal Whaling in Japan: Report of an International Workshop*, Edmonton, Canada: Boreal Institute for Northern Studies.

Institute for Cetacean Research (1988c), "Whaling, Beliefs and Japanese World View," in *Small Type Coastal Whaling in Japan: Report of an International Workshop*, Edmonton, Canada: Boreal Institute for Northern Studies.

Institute for Cetacean Research (1988d), "Whale as Food in Contemporary Japan," in *Small Type Coastal Whaling in Japan: Report of an International Workshop*, Edmonton, Canada: Boreal Institute for Northern Studies.

Institute for Cetacean Research (1988e), "The Non-Commercial Distribution of Whale Meat," in *Small Type Coastal Whaling in Japan: Report of an International Workshop*, Edmonton, Canada: Boreal Institute for Northern Studies.

Institute for Cetacean Research (1988f), "Tourism," in *Small Type Coastal Whaling in Japan: Report of an International Workshop*, Edmonton, Canada: Boreal Institute for Northern Studies.

Institute for Cetacean Research (1988g), "Whale meat in Japanese Food Culture," in *Small Type Coastal Whaling in Japan: Report of an International Workshop*, Edmonton, Canada: Boreal Institute for Northern Studies.

Institute for Cetacean Research (1996a), *Papers on Japanese Small-Type Coastal Whaling Submitted by the Government of Japan to the International Whaling Commission, 1986–1995*, Institute for Cetacean Research (eds.), Tokyo: ICR.

Institute for Cetacean Research (1996b), "The Cultural Significance of Everyday Food Use," in *Papers on Japanese Small-Type Coastal Whaling Submitted by the Government of Japan to the International Whaling Commission, 1986–1995*, Institute for Cetacean Research (eds.), Tokyo: ICR.

Irwin, T., and Fine, G. (eds.) (1996), *Aristotle: Introductory Readings*, Indianapolis: Hackett Publishing.

Ishii, A., and Okubo, A. (2007), "An Alternative Explanation of Japan's Whaling Diplomacy in the Post-Moratorium Era," *Journal of Wildlife Law and Policy*, vol. 10, pp. 55–87.

Isihara, A., and Yoshii, J. (2000), *A Survey of the Commercial Trade in Whale Meat Products*, TRAFFIC Report, www.traffic.org, accessed 04/09/2004.

IWC (1950–2004), *Report(s) of the International Whaling Commission*, Cambridge: IWC.

IWC/53/18 (2001), A South Pacific Whale Sanctuary, submitted by the Governments of Australia and New Zealand.

IWC/59/7 (2007), "Conference for the Normalization of the International Whaling Commission: Chair's Summary."

Jackson, G. (1978), *The British Whaling Trade*, London: Adam and Charles Black.

Jaeger, H. M. (2007), "'Global Civil Society' and the Political Depoliticisation of Global Governance," *International Political Sociology*, vol. 1, no. 3, pp. 257–277.

Jameson, A. (1996), "The Shaping of the Global Agenda: The Role of Non-Governmental Organisations," in *Risk, Environment and Modernity: Towards a New Ecology*, Lash, S., Szerszynski, B., and Wynne, B. (eds.), London, Thousand Oaks, and New Delhi: Sage.

Jameson, A., and Eyerman, R. (1989), "Environmental Knowledge as Organizational Weapon: The Case of Greenpeace," *Social Science Information*, vol. 28, no. 1, pp. 99–119.

Jameson, F. (1984), "Postmodernism or the Cultural Logic of Late Capitalism," *New Left Review*, vol. 46, pp. 53–92.

Jameson, F. (1991), *Postmodernism or the Cultural Logic of Late Capitalism*, Durham: Duke University Press.

Jarowski, A., and Coupland, N. (1999), *The Discourse Reader*, London and New York: Routledge.

Jefferson, T. A., and Leatherwood, S. (1993), *Marine Mammals of the World*, IUCN/FAO/UNEP publications.

Jhally, S. (1989), "Advertising as Religion: The Dialectic of Technology and Magic," in Ien, A., and Jhally, S. (eds.), *Cultural Politics in Contemporary America*, London and New York: Routledge.

Joachim, J. (2003), "Framing Issues and Seizing Opportunities: The UN, NGOs and Women's Rights," *International Studies Quarterly*, vol. 47, pp. 247–274.

Johnson, L. E. (1991), *A Morally Deep World: An Essay on Moral Significance and Environmental Ethics*, Cambridge: Cambridge University Press.

Jordan, G., and Maloney, W. A. (1998), "Manipulating Membership: Supply-Side Influence on Group Size," *British Journal of Political Studies*, vol. 28, pp. 389–409.

Josselin, D., and Wallace, W. (2001), "Non-state Actors in World Politics: A Framework," in Josselin, D., and Wallace, W. (eds.), *Non-state Actors in World Politics*, New York and London: Palgrave.

Kalland, A. (1989), "The Spread of Whaling Culture in Japan," *Papers on Japanese Small-Type Coastal Whaling Submitted by the Government of Japan to the International Whaling Commission, 1986–1995*, Institute for Cetacean Research (eds.), Tokyo: ICR.

Kalland, A. (1994a), "A Concept in the Service of Imperialism? Aboriginal Subsistence Whaling," in *11 Essays on Whales and Man*, High North Alliance (eds.), Reine, Norway: HNA.

Kalland, A. (1994b), "Super Whale: The Use of Myths and Symbols in Environmentalism," in *11 Essays on Whales and Man*, High North Alliance (eds.), Reine, Norway: HNA.

Kalland, A. (1994c), "Whose Whale Is That? Diverting the Commodity Path," in *Elephants and Whales: Resources for Whom?*, Freeman, M. R., and Kreuter, U. P. (eds.), Geneva: Gordon and Breach Publishers.

Kalland, A. (1995), "Culture in Japanese Nature," in *Asian Perceptions of Nature: A Critical Approach*, Bruun, O., and Kalland, A. (eds.), Richmond, UK: Curzon.

Kalland, A. (1998), "The Anti-Whaling Campaigns and Japanese Responses," in *Japanese Position on the Anti-Whaling Campaign*, Tokyo: Institute for Cetacean Research.

Kalland, A., and Moeran, B. (1992), *Japanese Whaling: End of an Era?*, London: Curzon.

Kang, S., and Phipps, M. (2000), *A Survey of Whale Meat Markets along South Korea's Coast*, TRAFFIC Report, www.traffic.org, accessed 04/07/2003.

Kawharu, I. H. (ed.) (1989), *Waitangi: Maori and Pákehá Perspectives of the Treaty of Waitangi*, Oxford and Auckland: Oxford University Press.

Keck, M., and Sikkink, K. (1998), *Activists Beyond Borders: Advocacy Networks in International Politics*, Ithaca and London: Cornell University Press.

Keeley, J. (1990), "Towards a Foucauldian Analysis of International Regimes," *International Organization*, vol. 44, no. 1, pp. 83–105.

Kelley, L. (1995), "A Neo-Gramscian Approach to International Organisations: An Expanded Analysis of Current References to UN Development Activities," in

Boundaries in Question: New Directions in International Relations, MacMillan, J., and Linklater, A. (eds.), London and New York: Pinter.

Keohane, R. O. (1984), *After Hegemony: Cooperation and Discord in the World Political Economy*, Princeton, NJ: Princeton University Press.

Khuels, T. (1998), "Between Sovereignty and Environment: An Exploration of the Discourses of Government," in *The Greening of Sovereignty in World Politics*, Litfin, K. T. (ed.), Cambridge, MA: MIT Press.

Kinsella, S. (1995), "Cuties in Japan," in *Women, Media and Consumption in Japan*, Skov, L., and Moeran, B. (eds.), Honolulu: University of Hawaii Press and Richmond, UK: Curzon.

Klinowska, M. (1994), "Brains, Behaviour and Intelligence in Cetaceans," in *11 Essays on Whales and Man*, High North Alliance (eds.), Reine, Norway: HNA.

Klotz, A. (2002), "Transnational Activism and Global Transformations: The Anti-Apartheid and Abolitionist Experiences," *European Journal of International Relations*, vol. 8, no. 1, pp. 49–76.

Knobloch, F. (1996), *The Culture of Wilderness: Agriculture as Colonization in the American West*, Chapel Hill, NC, and London: University of North Carolina Press.

Komatsu, M., and Misaki, S. (2001), *The Truth Behind the Whaling Dispute*, Tokyo: ICR/MAFF.

Krasner, S. (ed.) (1983), *International Regimes*, Cornell, NY: Cornell University Press.

Kratochwil, F. (1989), *Rules, Norms, Decisions: On the Conditions of Practical Reasoning in International Relations*, Cambridge: Cambridge University Press.

Kuehls, T. (1996), *Beyond Sovereign Territory: The Space of Ecopolitics*, Minneapolis: University of Minnesota Press (Borderlines).

Kuehls, T. (1998), "Between Sovereignty and Environment: An Exploration of the Discourse of Government," in *The Greening of Sovereignty in World Politics*, K. T. Litfin ed., Cambridge, MA: MIT Press.

Kutting, G. (2000), *Environment, Society and International Relations: Towards More Effective International Agreements*, London and New York: Routledge.

Laclau, E. (ed.) (1994), *The Making of Political Identities*, London: Verso.

Laclau, E., and Mouffe, C. (1985), *Hegemony and Socialist Strategy: Towards a Radical Democratic Politics*, London and New York: Verso.

Laffey, M. (1999), "Adding an Asian Strand: Neoliberalism and the Politics of Culture in New Zealand, 1984–97," in *Cultures of Insecurity: States, Communities and Danger*, Weldes, J., Laffey, M., Duvall, R., and Gusterson, H. (eds.), Minneapolis: University of Minnesota Press.

Lanfant, M., and Graeburn, N. (1994), "International Tourism Reconsidered: The Principle of the Alternative," in *Tourism Alternatives: Potentials and Problems in the Development of Tourism*, Smith, V. L., and Eadington, W. R. (eds.), Chichester: Wiley.

Lapid, Y. (1989), "The Third Debate: On the Prospects of International Relations Theory in a Post-Positivist Era," *International Studies Quarterly*, vol. 33, no. 3, pp. 235–254.

Larsen, H. (1999), "British and Danish European Policies in the 1990s: A Discourse Approach," *European Journal of International Relations*, vol. 5, no. 4, pp. 451–483.

Lash, S., Szerszynski, B., and Wynne, B. (eds.) (1996), *Risk, Environment and Modernity: Towards a New Ecology*, London, Thousand Oaks, and New Delhi: Sage.

Latour, B. (1986), *Laboratory Facts: The Construction of Scientific Facts*, Princeton, NJ: Princeton University Press.

Lee, D. (1992), *Competing Discourses: Perspective and Ideology in Language*, London and New York: Longman.

Lévi-Strauss, C. (1968), *Structural Anthropology, Vol. 1*, London: Penguin.

Lewis, M., and Shukovsky, P. (2007), Tribe Vows Prosecution for the Killing of Whales, *Seattle Post-Intelligencer*, 10/09/2007.

Lidskog, R., and Sundqvist, G. (2002), "The Role of Science in Environmental Regimes: The Case of LRTAP," *European Journal of International Relations*, vol. 8, no. 1, pp. 77–101.

Lipschutz, R. (1992), "Reconstructing World Politics: The Emergence of Global Civil Society," *Millennium*, vol. 21, no. 3, pp. 389–420.

Lipschutz, R. D., and Conca, K. (1993), *The State and Social Power in Global Environmental Politics*, New York: Columbia University Press.

Litfin, K. (1994), *Ozone Discourses: Science and Politics in Global Environmental Cooperation*, New York: Columbia University Press.

Litfin, K. (ed.) (1998), *The Greening of Sovereignty in World Politics*, Cambridge, MA: MIT Press.

Longworth, J. W. (1983), *Beef in Japan: Politics, Production, Marketing and Trade*, St. Lucia/Brisbane: University of Queensland Press.

Luhrmann, T. L. (1993), "The Resurgence of Romanticism: Contemporary Neopaganism, Feminist Spirituality and the Divinity of Nature," in *Environmentalism: The View from Anthropology*, Milton, K. (ed.), London and New York: Routledge.

Luke, T. (1995), "Sustainable Development as a Power/Knowledge System: The Problem of 'Governmentality,'" in *Greening Environmetnal Policy: The Politics of a Sustainable Future*, Fisher, F., and Black, M. (eds.), London: Paul Chapman.

Luke, T. (1999), "Environmentality as Green Governmentality," in *Discourses of the Environment*, Darier, E. (ed.), Oxford: Blackwell.

Lukes, S. (2005), *Power: A Radical View*, London: Palgrave Macmillan.

Lynn Doty, R. (1996), *Imperial Encounters: The Politics of Representation in North–South Relations*, Minneapolis: University of Minnesota Press (Borderlines).

MacCannell, D. (1999), *The Tourist: New Theory of the Leisure Class* (3rd ed.), Berkeley, CA: University of California Press.

Mackintosh, N. A. (1965), *The Stocks of Whales*, The Buckland Foundation, London: Coward and Gerrish Ltd.

MacMillan, J., and Linklater, A. (eds.) (1995), *Boundaries in Question: New Directions in International Relations*, London and New York: Pinter.

Magnusson, W., and Shaw, K. (eds.) (2003), *A Political Space: Reading the Local Through Clayoquot Sound*, Minneapolis: University of Minnesota Press.

Marx, K. (1990, English ed.), *Capital. A Critique of Political Economy, Vol. 1*, London: Penguin Books/New Left Review.

Matthews, L. H. (ed.) (1975), *The Whale*, Gothernburg, Sweden: Ab Nordbok.

Maxwell, S. (1996), "Food Security: A Post-modern Perspective," *Food Policy*, vol. 21, no. 2, 155–170.

Mazzanti, M. (2001), "The Role of Economics in Global Management of Whales: Reforming or Re-founding the IWC?," *Ecological Economics*, vol. 36, pp. 205–221.

McCormick, J. (1989), *The Global Environmental Movement: Reclaiming Paradise*, London: Behaven Press.

McDorman, T. (1995), "Protecting International Marine Living Resources with Trade Embargoes: GATT and International Reaction to US Practices," in *Additional Essays on Whales and Man*, High North Alliance (eds.), Reine, Norway: HNA.

Melville, H. (1851), *Moby-Dick*, Harmondsworth: Penguin.

Mercer, J. (2005), "Rationality and Psychology in International Politics," *International Organization*, vol. 59, pp. 77–106.

M'Gonigle, M. (1980), "The 'Economizing' of Ecology: Why Big, Rare Whales Still Die," *Ecology Law Quarterly*, vol. 9, no. 1, pp. 119–237.

Milliken, J. (1999), "The Study of Discourse in International Relations: A Critique of Research and Methods," *European Journal of International Relations*, vol. 5, no. 2, pp. 225–254.

Mills, S. (1997), *Discourse*, London and New York: Routledge.

Millstone, E., and Lang, T. (2003), *The Atlas of Food*, London: Earthscan.

Milton, K. (ed.) (1993a), *Environmentalism: The View from Anthropology*, London and New York: Routledge.

Milton, K. (1993b), "Environmentalism and Anthropology," in *Environmentalism: The View from Anthropology*, Milton, K. (ed.), London and New York: Routledge.

Misaki, S. (1994), "Whaling Controversy is the Name of the Game," in *Public Perception of Whaling*, Institute for Cetacean Research, Tokyo: ICR.

Misaki, S. (1996), *Whaling for the Twenty-First Century*, Tokyo: ICR.

Mitchell, R. B. (1998), "Forms of Discourse/Norms of Sovereignty: Interests, Science, and Morality in the Regulation of Whaling," in *The Greening of Sovereignty in World Politics*, Litfin, K. (ed.), Cambridge, MA, and London: MIT Press.

Mitchell, R. B. (2006), "International Environment," in *Handbook of International Relations*, Carslnaes, W., Risse, T., and Simmons, B. A. (eds.), London, Thousand Oaks, and New Delhi: Sage.

Moeran, B. (1992), "The Cultural Construction of Value: 'Subsistence,' 'Commercial' and Other Terms in the Debate about Whaling," *Maritime Anthropological Studies*, vol. 5, pp. 1–15.

Mofson, P. (2000), "Zimbabwe and CITES: Influencing the International Regime," in *Endangered Species, Threatened Convention: The Past, Present and Future of CITES*, Hutton, J., and Dickson, B. (eds.), London: Earthscan.

Morton, H. (1982), *The Whale's Wake*, Honolulu: University of Hawaii Press.

Mottier, V. (2002), "Discourse Analysis and the Politics of Identity/Difference," *European Political Science*, vol. 2, no. 1, pp. 57–60.

Moyle, B. J., and Evans, M. (2001), "A Bioeconomic and Socio-Economic Analysis of Whale-Watching, With Attention Given to Associated Direct and Indirect Costs," paper presented to the IWC Scientific Committee, SC/53/WW8.

Murphy, R. (1994), *Rationality and Nature: A Sociological Inquiry into a Changing Relationship*, Boulder, San Francisco, and Oxford: Westview Press.

Nadelmann, E. A. (1990), "Global Prohibition Regimes: The Evolution of Norms in International Society," *International Organization*, vol. 44, no. 4, pp. 479–526.

Nash, R. (1973), *Wilderness and the American Mind*, New Haven, CT: Yale University Press.

National Food and Nutrition Committee of Tonga (1999), "Policy Paper Addressing the Health Risks of Consumption of Mutton Flaps in Tonga," in *Issues Related to Indigenous Whaling: Tonga*, Freeman, M. (ed.), Brentwood Bay, Canada: World Council of Whalers.

Neumann, I. B. (1996), "Self and Other in International Relations," in *European Journal of International Relations*, vol. 2, no. 2, pp. 139–174.

O'Connell, D. P. (1982), *The International Law of the Seas, Vol. 1*, Oxford: Clarendon Press.

Onuf, N. G. (1989), *World of Our Making: Rules and Rule in Social Theory and International Relations*, Columbia, SC: University of South Carolina Press.

O'Regan, T. (1996), Letter to Mr Hubert Markishtum, head of the Makah Tribal Council, 21/06/1996.

Organisation for Economic Co-operation and Development (2001), *Review of Fisheries in OECD Countries: Policies and Summary Statistics*, Paris: OECD Publications.

Organisation for Economic Co-operation and Development (2002a), *Agricultural Policies in OECD Countries: Monitoring and Evaluation*, Paris: OECD Publications.

Organisation for Economic Co-operation and Development (2002b), *OECD Agricultural Outlook 2002–2007*, Paris: OECD Publications.

Paddock, R. C. (2001), "A Vexing Virulent, Varmint," *Los Angeles Times*, 25/05/01.

Parsons, C. (2007), *How to Map Arguments in Political Science*, Oxford: Oxford University Press.

Paterson, M. (1994), *Exploring the Climate Change Convention: Global Warming and IR Theory*, unpublished PhD dissertation, University of Essex.

Paterson, M. (2001), *Understanding Global Environmental Politics: Domination, Accumulation, Resistance*, Basingstoke, UK: Palgrave.

Pearce, F. (1991), *Green Warriors: The People and the Politics behind the Environmental Revolution*, London: Bodley Head.

Pentinnen, E. (2000), "Capitalism as a System of Global Power," in *Power in Contemporary Politics*, Goverde, H., Cerny, P. H., Haugaard, M., and Lentner, H. (eds.), London, Thousand Oaks, and New Dehli: Sage.

Perrez, F. X. (1996), "The Relationship between 'Permanent Sovereignty' and the Obligation Not to Cause Transboundary Environmental Damage," *Environmental Law*, vol. 26, no. 4, pp. 1187–1212.

Persaud, R. B. (2001), *Counter-Hegemony and Foreign Policy: The Dialectics of Marginalized and Global Forces in Jamaica*, Albany: State University of New York Press.

Peterson, M. J. (1992), "Whalers, Cetologists, Environmentalists and the International Management of Whaling," *International Organization*, vol. 46, no. 1, pp. 147–186.

Porter, G., Welsh Brown, J., and Chasek, P. S. (2000), *Global Environmental Politics*, 3rd ed., Boulder, CO, and Oxford: Westview Press.

Porterfield, H., and Denn, R. (1999), At Sea and Ashore, Insults Fly in Furor Over the Hunt, *Seattle Post Intelligence*, 19/05/09.

Poster, M. (1997), *Cultural History + Postmodernity: Disciplinary Readings and Challenges*, New York: Columbia University Press.

Pouillot, V. (2007), "'Sobjectivism': Towards a Constructivist Methodology," *International Studies Quarterly*, vol. 51, no. 2, pp. 359–384.

Price, R., and Reus-Smit, C. (1998), "Dangerous Liasons? Critical International Theory and Constructivism," *European Journal of International Relations*, vol. 4, no. 3, pp. 259–294.

Princen, T., Maniantes, M., and Conca, K. (eds.) (2002), *Confronting Consumption*, Cambridge, MA: MIT Press.

Raysmaker, C. (2001), *Monitoring Progress in Norway's Development of a DNA Register as Part of Its Domestic Management System for Whale Meat, Investigating Local Whale Meat Trade, and Investigating Reports of Illegal Trade in Blubber*, TRAFFIC Report, www.traffic.org.

Redclift, M. (2005), "Sustainable Development (1987–2005): An Oxymoron Comes of Age," *Sustainable Development*, vol. 13, pp. 212–227.

Redclift, M., and Benton, T. (eds.) (1994), *Social Theory and the Global Environment*, London and New York: Routledge.

Redclift, M., and Woodgate, G. (eds.) (2000), *The International Handbook of Environmental Sociology*, Cheltenham, and Northhampton, MA: Edward Elgar.

Rifkin, J. (1992), *Beyond Beef, the Rise and Fall of the Cattle Culture*, London: Thorsons/ HarperCollins.

Ris, M. (1994), "Why Look at Whales? Reflections on the Meaning of Whale Watching," in *11 Essays on Whales and Man*, High North Alliance (eds.), Reine, Norway: HNA.

Risse, T. (2006), "Transnational Actors and World Politics," in *Handbook of International Relations*, Carslnaes, W., Risse, T., and Simmons, B. A. (eds.), London, Thousand Oaks, and New Delhi: Sage.

Rittberger, V. (1991), "Regime Theory and International Environmental Management," in *The International Politics of the Environment: Actors, Interests, Institutions*, Hurrell, A., and Kingsbury, B. (eds.), Oxford: Clarendon Press.

Rose, C. (1993), "Beyond the Struggle for Proof: Factors Changing the Environmental Movement," *Environmental Values*, vol. 2, pp. 285–298.

Rose, G. (1996), "International Law and the Status of Cetaceans," in *The Conservation of Whales and Dolphins: Science and Practice*, Simmonds, M. P., and Hutchinson, J. D. (eds.), Chichester: Wiley.

Rosow, S. (1994), "Nature, Need, and the Human World: 'Commercial Society' and the Construction of the World Economy," in *The Global Economy as Political Space*, Rosow, S., Inayatullah, N., and Rupert, M. (eds.), Boulder, CO, and London: Lynne Rienner.

Rubenstein, D. (1989), "Hate Boat: Greenpeace, National Identity, and Nuclear Criticism," in Der Derian, J., and Shapiro, M. (eds.), *International/Intertextual Relations: Postmodern Readings of World Politics*, Lexington: Lexington Books.

Ruggie, J. G. (1993), "Territoriality and Beyond: Problematizing Modernity in International Relations," *International Oganization*, vol. 47, no. 1, pp. 139–174.

Ruggie, J. G. (1998), *Constructing the World Polity: Essays on International Institutionalization*, London and New York: Routledge.

Runte, A. (1979), *National Parks: The American Experience*, Lincoln, NE: University of Nebraska Press.

Sachs, W. (2000), "Sustainable Development," in *The International Handbook of Environmental Sociology*, Redclift, M., and Woodgate, G. (eds.), Cheltenham, and Northhampton, MA: Edward Elgar.

Saco, D. (1999), "Colonizing Cyberspace: 'National Security' and the Internet," in Weldes, J., Laffey, M., Duvall, R., and Gusterson, H. (eds.), *Cultures of Insecurity: States, Communities and Danger*, Minneapolis: University of Minnesota Press.

Salmond, A. (1997), *Between Worlds—Early Exchanges between Maoris and Europeans 1773–1815*, Honolulu: University of Hawaii Press.

Sanderson, K. (1994), "*GRIND*—Ambiguity and Pressure to Conform: Faroese Whaling and the Anti-Whaling Protest," in *Elephants and Whales: Resources for Whom?*, Freeman, M. R., and Kreuter, U. P. (eds.), Geneva: Gordon and Breach Publishers.

Sandoe, P. (1994), "Do Whales Have Rights?," in *11 Essays on Whales and Man*, High North Alliance (eds.), Reine, Norway: HNA.

Sands, P. (1995), *Principles in International Environmental Law, Vol. I: Frameworks, Standards and Implementation*, Manchester: Manchester University Press.

Sands, P., Tarasofksy, R., and Weiss, M. (eds.) (1994), *Documents in International Environmental Law*, Manchester: Manchester University Press.

Saurin, J. (1996), International Relations, Social Ecology and the Globalisation of Environmental Changes," in *The Environment and International Relations*, Vogler, J., and Imber, M. (eds.), London and New York: Routledge.

Saxon, W, (2002), Christine Stevens, 84, a Friend to the Animals, *New York Times*, obituary, 15/10/02.

Scarff, J. E. (1977), "The International Management of Whales, Dolphins and Porpoise: An Interdisciplinary Assessment," *Ecology Law Quarterly*, vol. 6, no. 2 (part I), pp. 367–427, and vol. 3 (part II), pp. 574–638.

Schäffner, C., and Kelly-Holmes, H. (1996), *Discourse and Ideologies*, Clevedon, UK: Multilingual Matters Ltd.

Schiappa, E. (1996), "Towards a Pragmatic Approach to Definition: 'Wetlands' and the Politics of Meaning," in *Environmental Pragmatism*, Light, A., and Katz, E. (eds.), London and New York: Routledge.

Schiffrin, D. (1994), *Approaches to Discourse*, Oxford and Cambridge, MA: Blackwell.

Schmidt, V. (2008), "Discursive Institutionalism: The Explanatory Power of Ideas and Discourse," *Annual Review of Political Science*, vol. 11, pp. 303–326.

Schneider, V., and Pearce, D. (2004), "What Saved the Whales? An Economic Analysis of 20th Century Whaling," *Biodiversity and Conservation*, vol. 13, pp. 543–562.

Schwarz, U. (1991), "Geldmaschine Greenpeace," *Der Spiegel*, vol. 45, no. 38, pp. 84–105.

Schweder, T. (2000), "Distortion of Uncertainty in Science: Antarctic Fin Whales in the 1950s," *Journal of International Wildlife Law and Policy*, vol. 3, no. 1, pp. 73–92.

Schweder, T. (2001), "Protecting Whales by Distorting Uncertainty: Precautionary Mismanagement?," *Fisheries Research*, vol. 52, pp. 217–225.

Schweder, T., and Hagen, G. (1999), "A Note on the Cost of Instability in Whale Management," *Journal of Cetacean Research Management*, vol. 1, no. 2, pp. 137–140.

Searle, J. R. (1969), *Speech Acts: An Essay in the Philosophy of Language*, Cambridge: Cambridge University Press.

Seidel, G. (1985), "Political Discourse Analysis," in *Handbook of Discourse Analysis, Vol. IV*, Dijk, T. V. (ed.), London: Academic Press.

Shapiro, M. (2002), "Social Science, Geophilosophy and Inequality," *International Studies Review*, vol. 4, no. 3, pp. 25–46.

Shaw, G. C. (2003), "Clearcut Identities: Tracking Shapeshifters in Clayoquot Sound," in *A Political Space: Reading the Global through Clayoquot Sound*, Magnusson, W., and Shaw, K. (eds.), Minneapolis: University of Minnesota Press.

Shaw, K. (2002), "Indigeneity and the International," *Millennium*, vol. 31, no. 1, pp. 55–81.

Shaw, M. (1994), *Global Society and International Relations*, Cambridge: Polity Press and Malden, MA: Blackwell.

Shaw, M. (1997), *International Law*, 4th ed. Cambridge: Grotius Publications.

Shiraishi, M. (2003), "Characteristics of Whaling in Choshu and the Kayoi Whaling Company," in *Traditional Whaling Summit in Nagato*, Institute for Cetacean Research (ed.), Tokyo: ICR and Nagato: Nagato City.

Shiva, V. (1998), *Biopiracy: The Plunder of Nature and Knowledge*, Dartington: Green Books (The Gaia Foundation).

Sikkink, K. (1993), "The Power of Principled Ideas: Human Rights Policies in the United States and Western Europe," *Ideas and Foreign Policy: Beliefs, Institutions and Political Change*, Goldstein, J., and Keohane, R. O. (eds.), Ithaca and London: Cornell University Press.

Simmonds, M. P., and Hutchinson, J. D. (1996), *The Conservation of Whales and Dolphins: Science and Practice*, Chichester: Wiley.

Simmons, I. G. (1993), *Interpreting Nature: Cultural Constructions of the Environment*, London and New York: Routledge.

Simon, J. (2007), *Governing through Crime: How the War on Crime Transformed American Democracy and Created a Culture of Fear*, Oxford and New York: Oxford University Press.

Sklair, L. (1994), "Global Sociology and Global Environmental Change," in *Social Theory and the Global Environment*, Redclift, M., and Benton, T. (eds.), London and New York: Routledge.

Sklair, L. (1995), *Sociology of the Global System*, 2nd ed., Prentice Hall: Harvester Wheatsheaf.

Sklair, L. (2000), "Transnational Capitalist Class and the Discourse of Globalisation," *Cambridge Review of International Affairs*, vol. 14, no. 1, pp. 66–85.

Skov, L. "Environmentalism Seen through Japanese Women's Magazines," in *Women, Media and Consumption in Japan*, Skov, L., and Moeran, B. (eds.), Honolulu: University of Hawaii Press and Richmond, UK: Curzon.

Slijper, E. J. (1958), *Whales* (1979, 2nd ed.), London: Hutchinson.

Small, G. L. (1971), *The Blue Whale*, New York and London: Columbia University Press.

Smith, S. (1995), "The Self-Images of a Discipline: A Genealogy of International Relations Theory," in *International Relations Theory Today*, Booth, K., and Smith, S. (eds.), Cambridge: Polity Press.

Spencer, L., with Bollwerk, J., and Morais, R. C. (1991), "The Not So Peaceful World of Greenpeace," *Forbes Magazine*, November 11, pp. 174–180.

Spinoza, B. (1674), *Opera Posthuma*, Letter L (Letter to Jarig Jellis, June 2, 1674).

Stoett, P. J. (1997), *The International Politics of Whaling*, Vancouver: University of British Columbia Press.

Strand, C. (1996), "The Ethics and Politics of Whaling," in *The Conservation of Whales and Dolphins: Science and Practice*, Simmonds, M. P., and Hutchinson, J. D. (eds.), Chichester: Wiley.

Susskind, L. E. (1994), *Environmental Diplomacy: Negotiating More Effective Global Agreements*, Oxford: Oxford University Press.

Sutherland, J. (2000), "An Endangered Planet?" in *Contending Images of World Politics*, Fry, G., and O'Hagan, J. (eds.), London: McMillan Press and New York: St. Martin Press.

Sutton, P. W. (2000), *Explaining Environmentalism: In Search of a New Social Movement*, Aldershot, UK, and Burlington, VT: Ashgate.

Sylvester, C. (2002), *Feminist International Relations: An Unfinished Journey*, Cambridge, Cambridge University Press.

Taiaroa, A. (1999), "Sustainable Use and Indigenous Rights: A Maori Perspective," in *Issues Related to Indigenous Whaling: Tonga*, Freeman, M. (ed.), Brentwood Bay, Canada: World Council of Whalers.

Takahashi, J. (2003), "Whaling Culture in Contemporary Japan," in *Traditional Whaling Summit in Nagato*, Institute for Cetacean Research (ed.), Tokyo and Nagato: Institute for Cetacean Research and Nagato City.

Te Ohu Kai Moana/Treaty of Waitangi Fisheries Commission (1996), "Te Ohu Kai Moana and the Sustainable Use of Renewable Resources," Policy Paper submitted to the Ministry for Foreign Affairs and Trade, December.

Te Ohu Kai Moana/Treaty of Waitangi Fisheries Commission (1997), "Summary of a Hui a Iwi to Discuss the Sustainable Use of Whale Species," 11–12th June, Kaikoura, New Zealand.

Te Ohu Kai Moana/Treaty of Waitangi Fisheries Commission (1998), "Maori Rights to Whale Use," Paper presented at the World Council of Whalers Annual Assembly, Vancouver, BC.

Te Ohu Kai Moana/Treaty of Waitangi Fisheries Commission (1999), "Beached Whales as Food: Cetaceans and Maori Customary Use," Paper presented at the World Council of Whalers Annual Assembly, Iceland.

Te Ohu Kai Moana/Treaty of Waitangi Fisheries Commission (2000), "World Council of Whalers 3rd Annual Assembly Proceedings," 15th–19th November, Nelson, New Zealand.

Te Ohu Kai Moana/Treaty of Waitangi Fisheries Commission (2001), "New Zealand and the International Whaling Commission: A Maori View Point," Policy Paper submitted to the Ministry for Foreign Affairs and Trade, June.

The Christchurch Press, 18/11/2000.

Thomas, C. (ed.) (1994), *Rio: Unravelling the Consequences*, Ilford: Frank Cass.

Thompson, J. B. (1990), *Ideology and Modern Culture*, Cambridge: Polity Press.

Tolba, M. K., and Rummel-Bulska, I. (1998), *Global Environmental Diplomacy*, Cambridge, MA, and London: MIT Press.

Tomlinson, J. (1991), *Cultural Imperialism: A Critical Introduction*, London: Pinter.

Tonnessen, J. N., and Johnsen, A. O. (1982), *The History of Modern Whaling*, London: C. Hurst and Co. and Canberra: Australian National University Press.

Torfing, J. (1999), *New Theories of Discourse: Laclau, Mouffe and Zizek*, Oxford: Blackwell.

Torfing, J. (2002), "Discourse Analysis and the Post-structuralism of Laclau and Mouffe," *European Political Science*, vol. 2, no. 1, pp. 54–57.

Umezaki, Y. (1986), *Kujira-to Inbou*, http://luna.pos.to/whale, accessed 11/02/2004.

UN Conference on the Human Environment (1972), UN Doc. A/Conf.48/14, at 23 (June 5–16, 1972).

United Nations Environment Programme (1979), *Marine Mammals*, UNEP pamphlet.

U.S.C. (1972a), United States Code, Title 16, sections 1-1400, section 1361, "Declaration of Policy."

U.S.C. (1972b), United States Code, Title 16, sections 1-1400, section 1378 (a)(4).

U.S.C. (1975), United States Code, Title 22, para. 1978(a) (Supp. V).

U.S.C. (1976), United States Code, Title 16, para. 1383.

US Congressional Record (1971), Hearings before the Subcommittee on Fisheries and Wildlife Conservation, of the Committee on Merchant Marine and Fisheries, September 23, 1971, House Report 10420, vol. 117, part 34, 06/12/71.

Varner, G. E. (1994), "Environmental Law and the Eclipse of Land as Private Property," in *Ethics and Environmental Policy: Theory Meets Practice*, Ferré, F., and Hartel, P. (eds.), Athens, GA, and London: University of Georgia Press.

Vaughn Switzer, J. (2004), *Environmental Politics: Domestic and Global Dimensions*, Belmont, CA: Wadsworth/Thompson Learning.

Vogler, J., and Imber, M. (eds.) (1996), *The Environment and International Relations*, London and New York: Routledge.

Wacquant, L. (2001), "La pensée critique comme dissolvent de la doxa," *Revista de Filosofía*, vol. 26, no. 1, pp. 129–134.

Waever, O. (2004), "Discursive Approaches," in *European Integration Theory*, Wiener, A., and Diez, T. (eds.), Oxford: Oxford University Press.

Wales, C., and Mythen, G. (2002), "Risky Discourses: The Politics of GM Foods," *Environmental Politics*, vol. 11, no. 2, pp. 121–144.

Walker, R. B. J. (1993), *Inside/Outside: International Relations as Political Theory*, Cambridge and New York: Cambridge University Press.

Walloe, L. (1999), Statement by Norway at the 51st Annual Meeting of the IWC, http://www.highnorth.no/Library/Policies/National/lw-iwc-99.htm, accessed 07/11/07.

Wapner, P. (1996), *Environmental Activism and World Civic Politics*, Albany: State University of New York Press.

Watson, L. (1981), *Whales of the World*, London: Threshold Foundation.

Weldes, J. (1999), *Constructing National Interests: The US and the Cuban Missile Crisis*, Minneapolis: University of Minnesota Press.

Weldes, J., Laffey, M., Duvall, R., and Gusterson, H. (eds.) (1999), *Cultures of Insecurity: States, Communities and Danger*, Minneapolis: University of Minnesota Press.

Weldes, J., and Saco, D. (1996), "Making State Action Possible: The United States and the Discursive Construction of the 'Cuban Problem,' 1960–1994," *Millennium*, vol. 25, 1996, pp. 361–395.

Wendt, A. E. (1987), "The Agent-Structure Problem in International Relations Theory," *International Organization*, vol. 41, no. 3, pp. 335–370.

Wendt, A. E. (1992), "Anarchy is What States Make of It: The Social Construction of Power Politics," *International Organizations*, vol. 46, no. 2, pp. 391–425.

Wendt, A. (1999), *Social Theory of International Politics*, Cambridge: Cambridge University Press.

Whale Conservation Network (2001), Opening Statement, IWC 2001.

Wheeler, N. (2000), *Saving Strangers: Humanitarian Intervention in International Society*, Oxford: Oxford University Press.

Wight, C. (2004), "State Agency: Social Action Without Human Activity?," *Review of International Studies*, vol. 20, pp. 269–280.

Willets, P. (1982), *Pressure Groups in the Global System*, London: Frances Pinter.

Williams, G. (1999), *French Discourse Analysis: The Method of Post-structuralism*, London and New York: Routledge.

Williams, H. (1988), *Whale Nation*, London: Jonathan Cape.

Willis, R. G. (1990), *Signifying Animals: Human Meaning in the Natural World*, London, Boston, Wellinton, and Sydney: Unwin Hyman.

Wittgenstein, L. (1969), *Preliminary Studies for the "Philosophical Investigations," Generally Known as the Blue and Brown Books*, Oxford: Blackwell.

Wodak, R., and Meyer, M. (2001), *Methods of Critical Discourse Analysis*, London: Sage.

World Commission on Environment and Development (1987), *Our Common Future*, Oxford and New York: Oxford University Press.

World Council of Whalers (1998a), "WCW Resolution Entered into Congressional Record," Press Release, 11/06/1998.

World Council of Whalers (1998b), *Facts and Issues*, http://www.worldcouncilofwhalers.com/faqsframe.htm, accessed 11/11/2004.

World Council of Whalers (2002), "World Council Says No to Large Scale Whaling," Press Release, 30/06/2000, http://www.worldwhalers.com, accessed 12/11/2004.

World Council of Whalers (2004), "WCW Profile," http://www.worldwhalers.com/wcwprofile/history.htm, accessed 12/11/2004.

World Council of Whalers (1999), *Issues Related to Indigenous Whaling: Tonga*, Freeman, M. (ed.), Brentwood Bay, Canada: World Council of Whalers.

Worster, D. (1977), *Nature's Economy: The Roots of Ecology*, San Francisco: Sierra Club Books.

Yearly, S. (1996), *Sociology, Environmentalism, Globalization*, London, Thousand Oaks, and New Delhi: Sage.

Young, O. (1989a), *International Cooperation: Building Regimes for Natural Resources and the Environment*, Cornell and London: Cornell University Press.

Young, O. (1989b), The Politics of International Regime Formation: Managing Natural Resources and the Environment," *International Organization*, vol. 43, no. 3, pp. 349–375.

Young, O. (1990), "Global Environmental Change and International Governance," *Millennium*, vol. 19, no. 3, pp. 337–347.

Young, O. (ed.) (1999a), *The Effectiveness of International Environmental Regimes: Causal Connections and Behavioural Mechanisms*, Cambridge, MA, and London: MIT Press.

Young, O. (1999b), *Governance in World Affairs*, Ithaca and London: Cornell University Press.

Zehfuss, M. (2001), "Constructivism and Identity: A Dangerous Liaison," *European Journal of International Relations*, vol. 7, no. 3, pp. 315–348.

List of Interviews

Bendiksen, Bjorn-Hugo, whaler. Lofoten Islands, Norway, July 31, 2003.

Bendiksen, Halvard, whaler. Lofoten Islands, Norway, July 30, 2003.

Briand, Frédéric, commissioner for Monaco. IWC annual conference 2003, Berlin, June 16, 2003.

Bridgewater, Peter, IWC chairman 1995–97, and vice-chairman, 1991–95. UNESCO, Paris, April 22, 2001.

Butterworth, Douglas, IWC SC member. London, October 13, 2003; followed up by e-mail correspondence, September 8, October 10, October 11, 2003, and January 1, 2004.

Clutton Brock, Tim, Large Animal Research Group, Cambridge University. Cambridge, April 9, 2002.

Cooke, Justin, IWC SC member. IWC annual conference 2003, Berlin, June 18th, 2003; followed up by e-mail correspondence, September 15, 2003.

Donovan, Greg, IWC scientific editor. IWC Cambridge, February 4, 2002.

Frovik, Rune, High North Alliance secretary. Lofoten Islands, Norway, July 27, 2003.

Gambell, Ray, former IWC secretary, 1976–2000. Cambridge, October 25, 2001; March 2002; August 2, 2002; and September 2003.

Garcia, Ismael, commissioner for Belize. IWC annual conference 2003, Berlin, June 16, 2003.

Graeburn, Nelson, anthropologist. University of California at Berkeley, May 2002.

Hagiwara, Mikiko, Safety First! IWC annual conference 2001, London, July 26, 2001.

Hasegawa, Ayako, WWF Japan. IWC annual conference 2001, London, July 26, 2001.

Hammond, Philip, IWC SC member and former SC chairman (until 1993). E-mail correspondence, September 3 and September 21, 2003.

Hastings, Jay, Japan Fisheries Association. IWC annual conference 2003, Berlin, June 17, 2003.

Holt, Sidney, former FAO, IWC SC member, former alternate commissioner for the Seychelles (1980–1982). IWC annual conference 2003, Berlin, June 18, 2003.

Kajiki, Tomoko, Safety First! IWC annual conference 2001, London, July 26.

Kerins, Sean, anthropologist. IWC annual conference 2001, London, July 24.

Kile, Nelson, commissioner for the Solomon Islands. IWC annual conference 2003, Berlin, June 17, 2003.

Kurasawa Nanami, Iruka and Kujira Action Network. IWC annual conference 2001, London, July 26, 2001.

Lichtervelde de, Alexander, commissioner for Belgium. IWC annual conference 2003, Berlin, June 16, 2003; followed up by a telephone interview, September 7, 2004.

Maack, Thilo, Greenpeace campaigner. IWC annual conference 2003, Berlin, June 15, 2003.

Macnow, Alan, Japan Whaling Association. IWC annual conference 2003, Berlin, June 18, 2003.

Mrosovsky, Nicholas, sea turtle conservation specialist. Cambridge, November 2001.

Naito, Mitsuru, WWF Japan. IWC annual conference 2001, London, July 26, 2001.

Olavsen, Jan-Odin, whaler. Lofoten Islands, Norway, July 31, 2003.

Palsboll, Per, IWC SC member. University of California at Berkeley, April 2002.

Pascal, Lloyd, commissioner for Dominica. IWC annual conference 2003, Berlin, June 18, 2003.

Phillips, Cassandra, WWF. IWC annual conference 2003, Berlin, June 16, 2003.

Simmons, William, anthropologist, Brown University. E-mail correspondence, June 17, 2002.

Index

Note: Organizations referred to in the text by both their acronyms and their full title are listed by the acronym, with the full title in parenthesis. Otherwise, organizations are listed as mentioned in the text. Notes are indicated by "n." after the page number, figures by "f."